T0093535

Optimization of Sustainable Enzymes Production

This book is designed as a reference book and presents a systematic approach to analyzing evolutionary and nature-inspired population-based search algorithms. Beginning with an introduction to optimization methods and algorithms and various enzymes, the book then moves on to provide a unified framework of process optimization for enzymes with various algorithms. The book presents current research on various applications of machine learning and discusses optimization techniques to solve real-life problems.

- The book compiles the different machine learning models for optimization of process parameters for the production of industrially important enzymes. The production and optimization of various enzymes produced by different microorganisms are elaborated in the book.
- It discusses the optimization methods that help minimize the error in developing patterns and classifications, which further helps improve prediction and decision-making.
- Covers the best-performing methods and approaches for the optimization of sustainable enzyme production with AI integration in a real-time environment.
- Featuring valuable insights, the book helps readers explore new avenues leading toward multidisciplinary research discussions.

The book is aimed primarily at advanced undergraduates and graduates studying machine learning, data science, and industrial biotechnology. Researchers and professionals will also find this book useful.

Optimization of Sustainable Enzymes Production

Artificial Intelligence and Machine Learning Techniques

Edited by
J. Satya Eswari
Nisha Suryawanshi

CRC Press
Taylor & Francis Group
Boca Raton London New York

CRC Press is an imprint of the
Taylor & Francis Group, an **informa** business

A CHAPMAN & HALL BOOK

First edition published 2023
by CRC Press
6000 Broken Sound Parkway NW, Suite 300, Boca Raton, FL 33487-2742

and by CRC Press
4 Park Square, Milton Park, Abingdon, Oxon, OX14 4RN

CRC Press is an imprint of Taylor & Francis Group, LLC

Library of Congress Cataloging-in-Publication Data
Names: Jujjavarapu, Satya Eswari, editor. | Suryawanshi, Nisha, editor.
Title: Optimization of sustainable enzymes production: artificial intelligence and machine learning techniques / edited by J. Satya Eswari, Nisha Suryawanshi.
Description: First edition. | Boca Raton: Chapman & Hall/CRC Press, 2023.
| Includes bibliographical references and index.
Identifiers: LCCN 2022025614 (print) | LCCN 2022025615 (ebook) | ISBN 9781032273372 (hardback) | ISBN 9781032273433 (paperback) | ISBN 9781003292333 (ebook)
Subjects: LCSH: Enzymes--Biotechnology. | Sustainable engineering.
| Mathematical optimization. | Artificial intelligence--Industrial applications.
| Machine learning--Industrial applications.
Classification: LCC TP248.65.E59 O67 2023 (print) | LCC TP248.65.E59 (ebook)
| DDC 628--dc23/eng/20220830
LC record available at https://lccn.loc.gov/2022025614
LC ebook record available at https://lccn.loc.gov/2022025615

ISBN: 978-1-032-27337-2 (hbk)
ISBN: 978-1-032-27343-3 (pbk)
ISBN: 978-1-003-29233-3 (ebk)

DOI: 10.1201/9781003292333

Typeset in Palatino
by SPi Technologies India Pvt Ltd (Straive)

Contents

Preface

Artificial intelligence has revolutionized the industrial procedures. Reduced optimization computational cost is one of the most common artificial intelligence applications. Several algorithms were introduced and tested based on their assumptions for solving issues, each with its own set of hypotheses. In this book, the principle of optimization is described and different optimization techniques such as genetic algorithm (GA), artificial neural network (ANN), particle swarm optimization (PSO), differential evolution (DE), and artificial bee colony (ABC) algorithm are used in some case studies for process optimization of enzyme development, and various models are completely developed and discussed. Machine learning is motivated by its ability to save resources, machining time, and energy while increasing yield in situations where traditional methods have reached their limits. This is a complicated task in which a huge number of controllable parameters have an impact on production in some way. Changing anywhere between 100 different control parameters to get the best combination of all the factors is required. The optimization problem is to identify the best combination of these parameters to optimize the production rate. This book takes a systematic method to examining population-based search algorithms that are evolutionary and nature-inspired.

The books on optimization currently available deal with machine learning and process optimization using algorithms or describe various algorithms for optimization. The applications of various machine learning models or algorithms, particularly for process optimization for the production of enzymes are not reported. This book aims to compile different machine learning models for optimization of process parameters for the production of industrially important enzymes. The production and optimization of various enzymes produced by different microorganisms are elaborated in the book.

The students and researchers in optimization, operations research, artificial intelligence, data mining, machine learning, computer science, and management sciences will see the pros and cons of a variety of algorithms through detailed examples and a comparison of algorithms.

Dr. J. Satya Eswari
Dr. Nisha Suryawanshi

Editors

Dr. J. Satya Eswari has been an assistant professor for more than 8 years at the Biotechnology Department of the National Institute of Technology (NIT), Raipur, India. She did her M.Tech in Biotechnology at the Indian Institute of Technology (IIT) Kharagpur and a Ph.D. at the IIT, Hyderabad, India. During her research career, she worked as a Scientist (Woman Scientist – Department of Science and Technology (DST)) in the Indian Institute of Chemical Technology (IICT), Hyderabad. She has published more than 60 SCI/Scopus research papers, 6 books, a few book chapters, and 40 international conference proceedings. Her research contributions have received wide global citations. She completed one DST woman scientist project (22 lakhs) and is currently handling one DST-Early career research project (43 lakhs) and one CCOST (4 lakhs). She has more than 7 years of teaching experience and 3 years of research experience. Dr. Eswari has been a guest editor for the *Indian Journal of Biochemistry and Biophysics* (SCI) and the *Journal of Chemical Technology and Biotechnology*. She has rigorously pursued her research in the areas of Environmental bioremediation, wastewater treatment, bioprocess, and product development and bioinformatics. She gained pioneering expertise in the application of mathematical and engineering tools to Biotechnological processes. She has received the IEI Young Engineer award, the Outstanding Woman by Venus International award, and the DK Best Faculty award. Dr. Eswari has already guided three Ph.D. students and is currently guiding three other Ph.D. students.

Dr. Nisha Suryawanshi is currently working as a guest faculty in the Department of Zoology at Government Arts and Commerce College, Sagar, Madhya Pradesh, India. She completed her Bachelor of Science (B.Sc.) in Biotechnology (Honours) from Guru Ghasidas Central University Bilaspur (Chhattisgarh, India), her Masters of Science (M.Sc.) in Biotechnology from Dr. Hari Singh Gour Central University, Sagar (Madhya Pradesh). She received a Doctor of Philosophy from the Department of Biotechnology, National Institute of Technology, Raipur (Chhattisgarh, India). She has ten publications in her research area in peer-reviewed SCI journals. During her Ph.D., she worked in the area of bioprocess and product development. She also has qualified national-level examinations CSIR-NET-JRF (Life science), GATE (Biotechnology), ICAR-NET (Agriculture Biotechnology), and the state-level examination MPSET.

Contributors

K. S. Anantharaju
Dayananda Sagar College of
    Engineering
Bangalore, India

P. V. Atheena
Manipal Institute of Technology
Manipal, India

Khashti Dasila
Kumaun University
Bhimtal, India

J. Satya Eswari
National Institute of Technology
Raipur, India

Karan Kumar
Indian Institute of Technology
Guwahati, India

Rajeev Kumar
Dayananda Sagar University
Bangalore, India

Vijayanand S. Moholkar
Indian Institute of Technology
Guwahati, India

Sunil S. More
Dayananda Sagar University
Bangalore, India

Uday Muddapur
KLE Tach University
Hubli, India

Ajay Nair
Dayananda Sagar University
Bangalore, India

Monalisa Padhan
Siddheswar College
Amarda Road, Balasore,
    Odisha, India

Sushri Priyadarshini Panda
Sambalpur University
Burla, India

Archana S. Rao
Dayananda Sagar University
Bangalore, India

Keyur Raval
National Institute of Technology
    Karnataka
Surathkal, India

Ritu Raval
Manipal Institute of Technology
Manipal, India

Heena Shah
Mandsaur University
Mandsaur, India

Dheeraj Shootha
Kumaun University
Bhimtal, India

C. Sowmya
Sapthagiri College of Engineering
Bangalore, India

Nisha Suryawanshi
Government Arts and Commerce
   College
Sagar, India

A. V. Narasimha Swamy
JNTUA
Anantapur, India

Pooja Thathola
Kumaun University
Bhimtal, India

S. M. Veena
Sapthagiri College of
   Engineering
Bangalore, India

1

Industrially Important Enzymes

A. V. Narasimha Swamy

JNTUA, Anantapur, India

CONTENTS

1.1 Introduction

There are diverse applications of enzymes in various fields and many more applications are being explored by scientists all over the world. Interestingly the present applications of enzymes include detergents like Surf excel, Ariel, Nirma, etc., and it became an essential constituent of detergents. If Biocon industries of Bangalore has ventured into producing enzymes on large scale in India, Novo enzymes Denmark has captured a major market segment in India.

Living organisms contain enzymes which are made of protein molecules. Enzymes act at a certain regulating rate at which biochemical reactions occur without being altered in the process. It is a characteristic feature of enzymes that they are highly specific in their action in catalyzing only one reaction step. Commercial production and utilization of enzymes are based on two facts: 1) Enzymes are produced by living cells. 2) Enzymes can exert their specific action independent of living cells. They play an important role in controlling major biochemical reactions which take place in the human body. The reactions which are performed by enzymes in various bioprocesses are economically viable and ecofriendly, when compared to reactions performed by chemical catalysts are costly and may not be always ecofriendly. The biochemical reactions that are carried out by enzymes are applied to different industries like food processing, leather processing, and textile processing. The various sectors like disease identification by means of diagnosis and monitoring incorporate enzymes of microbial origin.

The interior and exterior part of the cell encompasses the biochemical reactions that are carried out by enzymes. These enzymes are complex molecules of protein. The usage of enzymes in preserving food and beverage processing had been followed for ages. The enzymes can be derived either from microorganisms or from animal or plant resources and the enzymes that are derived from microorganisms exhibited better activity and stability.

However, the enzymes can be cultivated in a short time in huge quantities in novel bioreactors by fermentation and thus become an alternate source for enzyme production. The demand for enzymes is increasing in various sectors and this prompts the need of developing new strains of microorganisms. Certain cancers can be treated by employing enzymes of pancreatic origin (Wang et al., 2012). The defense mechanism of the body that acts against cancer contains proteolytic enzymes in the pancreas. Hence, these proteolytic enzymes may be used as anticancer agents in the future. The expression of gene action is primarily supported by the complex protein molecules derived from microbial origin. And the derived complex proteins participate in various biochemical reactions. The lower energy pathways for the reactants and products are provided by complex protein molecules that participate in biochemical reactions as biocatalysts.

The enzymes which are presently available on market are those which act on amylose, lipids, cellulose, and proteins. As of now, nearly 5% of about 4000 enzymes known are available in the market. DuPont, Roche, and Novozymes share the major enzyme market in the world now.

1.2 Structure

The protein structure is shown in Figure 1.1

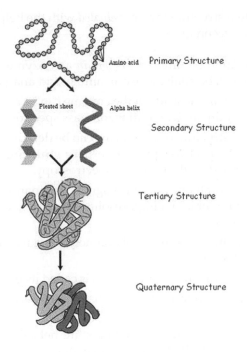

FIGURE 1.1
Structure of protein.

The peptide bond *exists* between the amino group (-NH$_2$) of one amino acid and the carboxyl group (-COOH) of another.

The peptide bond I protein is shown in Figure 1.2.

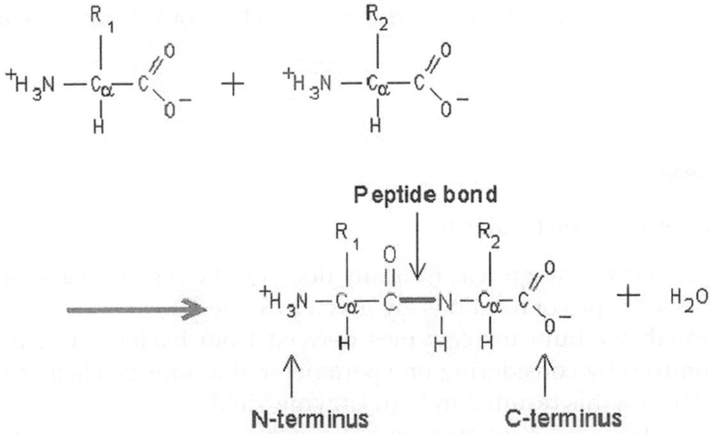

FIGURE 1.2
Peptide bond in protein.

Analysis of protein structure can be *evaluated* with the help of the following advanced analytical techniques.

- The determination of amino acids *within* an enzyme and the molar ratios of each can be analyzed by an amino acid analyzer.
- The sequence of amino acids in the enzyme can be analyzed by peptide mapping, Edman degradation, or mass spectroscopy.
- The secondary structure of an enzyme can be determined by circular dichroism (CD) spectroscopy. The tertiary structure of an enzyme can be determined by fluorescence spectroscopy.
- X-ray crystallography or nuclear magnetic resonance (NMR) analysis can be used to obtain a high-resolution analysis of the 3D structure of an enzyme.
- Enzyme structure has four levels namely primary, secondary, tertiary, and quaternary.
- The amino acid sequence of an enzyme is called its primary structure
- The interaction of amino acids in a chain is the secondary structure of enzymes.
- The two types of secondary structures are helical (called αhelices) and pleated sheets (called βpleated sheets).
- The arrangement of amino acids in three-dimensional spaces is the Tertiary structure. Quaternary structure refers to the interaction between protein subunits.
- Analysis of enzyme structure can be done with the help of advanced analytical techniques/equipment such as amino acid analyzer, peptide mapping, Edman degradation, mass spectroscopy, circular dichroism spectroscopy, fluorescence spectroscopy, X-ray crystallography, NMR.

1.3 Enzyme Classification

The enzyme classification was basically done on the type of reaction carried out. The seven types of reactions are given in Table 1.1.

The growth medium for enzymes derived from bacteria and fungi has been optimized by considering one parameter at a time by Hajji et al. (Hajji et al., 2008) And this resulted in high enzyme yield.

The evaluation and regulation of the process parameters were not easily achieved (Wang et al., 2012). Interestingly, solid-state fermentation as well as submerged fermentation was used successfully for the production of a few

TABLE 1.1

Classification of Enzymes

Classification	Reaction	Description
Oxidoreductases – EC1	$A_r + B_o \rightleftarrows B_r + A_o$	Catalyzes redox reaction
Transferases – EC2	$A + B + C \rightarrow A - B - C$	Catalyze transfer of groups
Hydrolases – EC3	$A + B + H_2O \rightarrow AH + BOH$	Promotes hydrolysis
Lyases – EC4	$A - B \rightleftarrows A + B$	Catalyzes removal of groups or reverse reaction
Isomerases – EC5	$A - B - C \rightleftarrows A - C - B$	Catalyzes conversion of isomers
Lygases – EC6	$A - B + ATP \rightarrow A - B + ADP + P_i$	Catalyzes synthesis of two molecular substrates into one with energy release
Translocases – EC7		Catalyzes movements of ions

Note: r = reduction, o = oxidation

enzymes. Submerged fermentation and solid-state fermentation by using peels of orange were found to be advantageous for producing the enzyme Invertase. Alcoholic beverages like ethanol can be produced by enzymes (Lincoln & More, 2017). *Aspergillus* species have the capability to utilize agricultural substrates containing lignocellulosic material to produce enzyme-like cellulase (Pachauri et al., 2018). The batch mode and fed-batch mode bioreactors have been used during microbial fermentation to produce industrial enzymes. Fed-batch bioreactors were employed to produce protease enzymes by *Alcaligenes* bacteria for their usage in detergents (Aishwarya et al., 2013). The microbial production of industrial enzymes has been optimized by using different statistical methods. Plackett–Burman method. was used to optimize, process parameters to enhance tannase enzyme production by *Aspergillus* species (Xiao et al., 2015; Souza et al., 2015). The substance Tween 80 which is present in the medium may also stimulate bacteria like *Bacillus* for enzyme production. The surfactants modify the plasma membrane to facilitate compound uptake to enhance laccase enzyme release (Niyonzima & More, 2014a). The usage of surfactants like triton in fermentation medium has increased extracellular enzyme secretion by *Penicillium* species. The aromatic compounds also act as inducers and contribute to stimulating laccase enzyme production by mushrooms (Usha et al., 2014). Environmental pollution can be drastically reduced by using cheap substrates like agricultural residues. The cost of production of microbial can be minimized by using inexpensive substrates (Niyonzima & More, 2019). Different raw materials like chicken feather are used instead of carbon and nitrogen source for keratinase production in large amounts. The huge demand for industrial enzymes leads to the coproduction of enzymes in the same medium by microorganisms to make the process inexpensive. Industrial enzymes can be produced simultaneously under similar conditions by bacteria or fungi to make the process inexpensive. However, stability is observed in the enzymes that are produced simultaneously in the same medium. In order to remove stains the enzymes like lipase, protease and cellulase are added to detergents during manufacture. The proteolysis of lipase, amylase, and cellulase, by the protease is averted due to the simultaneous cultivation of enzymes in the same medium. The proteolysis by amylase and lipase is not found when simultaneously lipase and amylase were produced by *Bacillus* species in the same fermentation medium. The protease hydrolysis resistant enzymes like cellulase and lipase have been found to be compatible in the formulation of detergents. The alkaline amylase of *Bacillus megaterium B69* was also not hydrolyzed by the alkaline protease when collaterally produced together. The production of industrial enzymes can be increased by inclosing agricultural byproducts as cheap raw materials. The enzymes amylase and protease can be produced by solid-state fermentation by using inexpensive substrates. The deoiled cake of muster seed also can be used as a cheap substrate for amylase and protease enzyme production economically. Likewise, the agricultural-based cheap raw material was employed for the production of lipase and amylase by *Bacillus* species. The fermentation cost of industrial enzyme production can be reduced by using cheap raw materials like agricultural residue. The various process parameters, such as agitation speed, time of incubation, the concentration of seed, pH, micronutrients, and macronutrients, contribute to industrial enzyme production. The yield of industrial enzymes can be increased by optimizing process parameters. The process parameters are usually optimized one factor each time, holding all other factors unchanged and the optimized condition/factor is considered in further experiments in sequential order. The techno-economically viable fermentation for enzyme production can be achieved by optimizing different parameters relating to nutrition factors, fermentation conditions, and process parameters.

1.3.1 Microbial Enzymes

The partial list of microbial enzymes is given in Table 1.2.

TABLE 1.2

Part List of Microbial Enzymes

S. No.	Enzyme	Microorganism
1	α-amylase	*Aspergillus niger, Bacillus subtilis*
2	Protease	*A. niger, B. subtilis*
3	Lipase	*A. niger, Candida cylindracea*
4	Cellulase	*A. niger, Trichoderma viride*
5	Invertase	*Saccharomyces cerevisiae*
6	Pectinase	*Conlosthyrium diplodiella*
7	Glucose isomerase	*Streptomyces* sp.

1.4 Industrial Enzyme Applications

The enzyme's applications include personal care sector, medical sector, food processing, healthcare, and detergent sector.

Table 1.3 depicts enzyme applications based on different industrial sectors and technical benefits in various fields.

Pulp and paper, and biofuels sector are a few to mention regarding the application of the technical enzyme in bulk. By looking at the present trend, it may be predicted that the technical enzymes market will increase to reach about $2.0 billion in 2025.

The enzymes that abundantly exist in the market include, amylose degrading, protein degrading, and cellulose-degrading enzymes. Amylose degrading enzymes have huge applications in the enzyme industry mainly due to the abundance of starch and are useful in the conversion of starch to sugar syrups. The enzyme applications can be further widened by developing amylase enzymes which have thermophilic, thermo–tolerant, and pH-tolerant characteristics. These thermophilic amylase enzymes would be directed to improve the gelatinization of starch, reduce the viscosity of the medium, and speed up biocatalysis. The most thermo-stable α-amylase presently used in industrial processes is obtained from *B. licheniformis*. It can sustain at a temperature of 90°C, for a long time.

TABLE 1.3

Partial List of Enzyme Applications

Application Field	Enzyme	Technical Benefits
Pulp and paper	Amylase	Cleaving starch molecules to reduce the viscosity for surface sizing in coatings, but not used for dry strength agent additive.
	Lipase	Deinking and control pitch in the pulping process
	Cellulases	Improving softness by hydrolyzing cellulose in fibers, creating weak spots in fibers, and making fibers flexible.
	Mannanases	Degrading residual glucomannan to increase brightness
	Laccases	Bleaching to improve brightness
	β xylanases	Enhancing the pulp bleaching process efficiently
Textile Industry	Amylases	Desizing efficiently without harmful effects to fabric
	Cellulases	Removing fuzz and microfibers to give fabric a smooth and glossier effect
	Pectinases	Losing indigo dye on denim to give a slightly worn look
	Proteases	Hydrolysis of protein molecules.

The applications of Lipases as biocatalysts have increased tremendously in the present due to their high stability in organic solvents. Enzymes' participation in esterification, transesterification, aminolysis, and oximolysis reactions, would be desirable for green synthesis. It is interesting to note that the usage of enzymes like lipase is not predominant due to its low stability in process conditions and low activity on certain substrates. Therefore, it became essential to solving the problem of low activity and it may be solved by adopting two ways. i) to develop highly powerful enzymes from extremophilic microorganisms. which can withstand high temperature, pH, and concentration of salts. The method that evolved is systems microbiology and it has become a useful tool.

Cellulose degrading enzymes are largely applied in textile units for a long time and gained significance due to their capability to degrade lignocellulosic materials. The enzymatic conversion of lignocellulosic waste biomass to bioethanol has become a costly proposition due to a lack of feasible technology. This lead to cost reduction of Cellulase preparations by i) changing enzyme production method using new strains and ii) enhancing the activity of the enzyme to reduce the quantity of enzyme required when compared to enzyme action by consortia. Some organizations developed novel cellulase enzymes with gene modifications and optimized enzyme release. An enzyme by the brand name Cellic was the outcome of Novozymes research which has the capability of converting agricultural biomass into bioethanol. This Cellic enzyme is better than the previously developed product by the same company and gives desired results even at low doses. This development is useful in bringing down the cost of bioethanol to about $2.0 per gallon and it is highly competitive when compared to present fuels.

1.4.1 Enzymes in Food Processing

The present lifestyle change in Europe, the USA, and the Pacific nations, toward high-quality, durable, and tasty foods lead to the evolution of new enzymes in food units. The present trend focuses on better quality in foods pertaining to natural aroma and pleasant taste. The enzyme application in the area of the food industry and beverage industry and future demand is expected to be $1.5 billion (Saibabu et al., 2013).

The applications of food-grade enzymes encompass the baking industry, fruit juice and cheese manufacturing, and wine making and brewing to enhance their flavor, texture, digestibility, and nutritional value. Enzymes are used either as processing aids or additives in food units. Examples of food additives are Lysozyme and Invertase. Enzymes used in food processing are typically sold as enzyme preparations, which contain not only the desired enzyme but also metabolites from the production strain and several added substances such as stabilizers. All these materials are expected to be safe as per good manufacturing practice (GMP).

The key item in evaluating enzyme preparation safety is the safety assessment of the production strain. Only about nine recombinant microorganisms are issued as Generally Recognized as Safe (GRAS) based on FDA regulations from a relatively small number of bacterial and fungal species primarily *Aspergillus oryzae*, *Aspergillus niger*, *Bacillus subtilis*, and *B. licheniformis*. In order to increase the enzyme production level, modifications including protease-deficient and sporulation-deficient were introduced to the wild-type host microorganisms.

1.4.2 Enzymes in Cosmetics

The predicted growth rate of enzymes by market experts in personal care is about 6% per year, due to developments in the enzyme production process. An example for technological progress is the usage of superoxide dismutase (SOD) Enzyme in personal care products for capturing free radicals to prevent skin problems. SOD and peroxidase may be mixed together in skin ointments as free radical removal to avert UV-prompted skin diseases. Also, protease enzyme is used in skin ointments for peeling off dead skin to evolve smooth skin. Further, it may be inevitable to stop skin irritations, which is the result of enzyme action on the skin.

However, many patents are being applied for enzyme usage in personal care products. Of course, the number of patent applications has increased since 2003 in the skin and cosmetic enzymes field.

The enzyme applications are given in Table 1.4.

TABLE 1.4

Enzyme Applications

S. No.	Enzyme	Application
1	Laccase	Hair dye
2	Lipase	Skin rash prevention
3	Endoglycosidase	Toothpaste/mouthwash for, whitening of teeth, odor removal
4	Papain	Toothpaste/mouthwash for whitening of teeth, odor removal
5	Catalase	Skin protection for whitening of teeth, odor removal

1.5 Commercially Important Enzymes

The main industrial enzymes may be divided into three categories: carbo-hydrate degrading, protein degrading, and lipid degrading. Amylases *come under* the carbohydrase group, *along with* cellulases, glucose isomerase, glu-cose oxidase, pectinases, xylanases, invertase, and galactosidase. The *amy-lose degrading* enzymes that exist in typically *in* markets are α-amylase and glucoamylase. A pertinent source of α-amylase is the bacteria *B. licheniformis*, whereas *A. oryzae* and *A. niger* are *frequently used* for glucoamylase indus-trial production. The *usual* industrial applications of α-amylase are in, bever-age (beer and distilled spirits), and *commercial* ethanol production; cleaning products; bakery products; textile processing; pulp and paper processing; animal feed; and digestive pharmaceuticals. Glucoamylase is basically used industrially for starch-based ethanol, with reducing market use in corn-steep liquor. α-amylase and glucoamylase put together accounted for 60% of the total carbohydrase.

The enzyme α-amylase is deployed in the industry due to its high thermal stability, and it is mostly obtained from microbial sources. The other sources of α-amylase are fungi and plants. The plant-based α-amylase sources are rice, corn, potato, and beetroots. For α-amylase, the annual growth rate of market demand in North America is expected to increase, by 2%, while the market growth rate in Asia is expected to increase by 6%.

The demand for glucoamylase was increased about two and a half fold in since last decade, whereas the main contribution is from glucoamylase, specifically for biofuels, as the main application, but for use in food and bev-erage processing, it was stable. The major industrial players of glucoamy-lase are Novozymes, DuPont, Amano Enzyme, AB Enzymes, Royal DSM, Verenium, Shandong Longda, VTR, SunHY, YSSH, and BSDZYME. The key markets are located in North America, China, Japan, India, Southeast Asia, and Europe.

1.6 Typical Enzyme Production Process

The industrial enzymes used in pharmaceutical industries are produced through bacterial and fungal fermentation. The bacterial and fungal strains are used because of their easy handling, fast growth rates, and ease of scale-up in large fermenters. The most predominantly used bacteria are *Escherichia coli*, *B. subtilis*, and the filamentous fungi that are used are *A. oryzae*, *A. niger*, etc.

1.6.1 Industrial Enzymes

Microbial industrial enzyme production is gaining a lot of importance. Microbial industrial enzyme production comprises important steps like isolation, screening, and identification of microorganisms which produce enzymes, and optimization of process optimization parameters. purification and characterization of purified enzymes.

Microorganisms are screened (Figure 1.3) *by following the conventional microbiological* methods The objective for isolating a suitable microbe attributes to (a) large quantity of enzyme release and less quantity of metabolites, (b) reduction of fermentation duration, and (c) cultivation of microbes in a cheaply available medium. Subsequently, the isolated microorganisms are optimized for their enzyme production optimized by strain improvement and media formulation. Microbial strains are developed by mutagenic chemicals and ultraviolet light.

The mutagenically treated microbial strains that produce enzymes are developed by spores and mycelia multiplication.

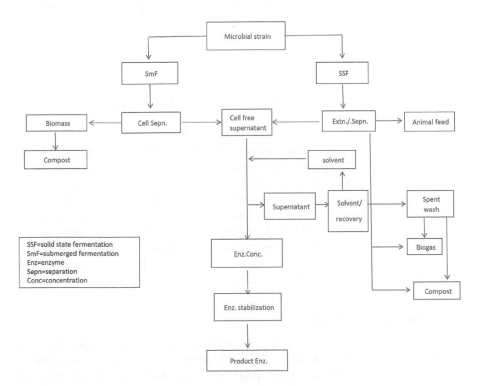

FIGURE 1.3
Microbial isolation.

The selection of microorganisms which produce enzymes in large quantities is of prime importance, in industrial enzyme production.

(1) Extracellular enzymes do not require cell disruption for their isolation, are preferred. (2) The microbial cultures that produce high-yielding enzymes are cultivated. (3) The strains that are stable regarding productivity are chosen. (4) The organism should be able to grow on cheap substrates. (5) The aspect is the ease of clarification of the culture liquor or extract. (6) The microbial Strain which does not produce toxic substances should be selected.

1.6.2 Medium Formulation and Preparation

The culture medium is formulated to achieve a large quantity production of enzymes and at the same time low growth rate of microbial strain. The medium that has a cheap source of carbon, nitrogen, amino acids, growth promoters, trace elements, and little amount of salts is considered. Optimum fermentation conditions like pH and temperature should be maintained during microbial strain development. The typical medium may contain (1) Carbohydrates like Molasses, barley, and corn and (2) proteins like soybean, peanut, and corn steep liquor.

1.6.3 Medium Sterilization

Medium is sterilization by batch and the continuous method is being considered for certain applications. Inoculation of sterile media for Fermentation would be the next step.

The enzyme production can be achieved by the surface culture technique or by the submerged culture technique. The surface culture technique is used for making certain fungal enzymes, like, amylase, protease, and pectinase.

The optimum fermentation conditions like pH, temperature, and oxygen are maintained. Foaming is controlled by adding oil during fermentation. The extracellular enzymes are produced, in a culture medium after an incubation period of 1–7 days. Most of enzymes are produced at the completion or during exponential phase. The metabolites about 10% produced in the fermented broth are removed during purification. The broth is maintained in the range of 3–5°C at the end of fermentation.

1.6.4 Purification of Enzymes

The steps involved in enzyme purification are i) vacuum evaporation for concentration, ii) filtration for polishing of enzyme solution, iii) addition of preservatives like calcium salts, sugar, alcohols, and sodium chloride, iv) precipitation of enzymes with acetone, alcohols, ammonium sulfate, etc., v) drying the precipitate by vacuum drying, and vi) packaging.

1.7 World Market

Globally, it is predicted that the enzyme market would rise about 7–8% annually since 2015. The innumerable applications of enzymes in industries and scientific research include textile industries, starch processing units, pharmaceutical units, and in bioprocess research. The huge demand for enzymes in food processing units and beverage units dictates the enzyme market in the world. The growth of enzyme industry is mostly governed by enzymes used in Bakeries and vegetable oil processing units. The growth pattern in Asia pacific may be marginal due to predominant development in Europe and United States. The quick growth of enzyme industry is evident from the huge demand of diagnostic enzyme kit, in medical care and health care sector. The detergent industry as well as cleaning enzymes manufacturing units has grown due to huge demand to decontaminate surfaces. Obviously enzyme production by bio route would be a right choice.

1.8 Summary

Various aspects like enzyme structure, classification, importance of commercial enzymes, and global market are given very briefly in this chapter and more information regarding enzymes is given in subsequent chapters.

Review Questions

1. What are enzymes?
2. In between which group's peptide bond is formed?
3. Explain about industrial applications of enzymes.
4. How enzymes are purified?
5. Briefly explain global scenario of enzymes.
6. Cite some applications of enzymes in cosmetics?

Bibliography

Abdel-Fattah YR, Soliman NA, Yousef SM, El-Helow ER (2012) Application of experimental designs to optimize medium composition for production of thermos table lipase/esterase by Geobacillus thermodenitrificans AZ1. *Journal, Genetic Engineering & Biotechnology* 10(2): 193–200.

Afsar A, Cetinkaya F (2008) Studies on the degreasing of skin by using enzyme in liming process. *Indian Journal of Chemical Technology*, 15(5): 507–510.

Aishwarya M, et al. (2013) Production, characterization and purification of alkaline protease from Alcaligenes sp. and its application in detergent industry. *Asian Journal of Pharmaceutical and Clinical Research* 6: 151–155.

Anisha GS, Sukumaran RK, Prema P (2008) Evaluation of a galactosidase biosynthesis by Streptomyces griseoloalbus in solid state fermentation using response surface methodology. *Letters in Applied Microbiology* 46:338–343.

Araujo R, Casal M, Cavaco-Paulo A (2008) Application of enzymes for textiles fibers processing. *Biocatalysis Biotechnology* 26: 332–349.

Banerjee A, Chisti Y, Banerjee UC (2004) Streptokinase – A clinically useful thrombolytic agent. *Biotechnology Advances* 22:287–307.

Bargagli E, Margollicci M, Nikiforakis N, et al. (2007) Chitotriosidase activity in the serum of patients with sarcoidosis and pulmonary tuberculosis. *Respiration* 74:548–552.

Berekaa MM, Zaghloul TI, Abdel-Fattah YR, Saeed HM, Sifour M (2009) Production of a novel glycerol-inducible lipase from thermophilic Geobacillus stearothermophilus strain-5. *World Journal of Microbiology and Biotechnology* 25(2): 287–294.

Bora L, Bora M (2012) Optimization of extracellular thermophilic highly alkaline lipase from thermophilic Bacillus sp.isolated from hot spring of Arunachal Pradesh, India. *Brazilian Journal of Microbiology* 43:30–42.

Brandelli A, Daroit DJ, Riffel A (2010) Biochemical features of microbial keratinases and their production and applications. *Applied Microbiology and Biotechnology* 85(6):1735–1750.

Cardenas JE, Alvarez MS, de Castro Alvarez JM, et al. (2001) Screening and catalytic activity in organic synthesis of novel fungal and yeast lipase. *Journal of Molecular Catalysis B: Enzymatic* 14:111–123.

Chen L, Shen Z, Wu J (2009) Expression, purification and in vitro antifungal activity of acidic mammalian chitinase against Candida albicans, Aspergillus fumigatus and Trichophyton rubrum strains. *Clinical and Experimental Dermatology* 34:55–60.

Choi JM, Han SS, Kim HS (2015) Industrial applications of enzyme biocatalysts: current status and future aspect. *Biotechnology Advances* 33: 1443–1454.

Dauter Z, Dauter M, Brzozowski AM et al (1999) X-ray structure of Novamyl, the five-domain 'maltogenic' a-amylase from Bacillus stearothermophilus: maltose and acarbose complexes at 1.7 A ° resolution. *Biochemistry* 38: 8385–8392.

De-Souza FR, Gutterres M (2012) Application of enzymes in leather processing: a comparison between chemical and coenzymatic processes. *Brazilian Journal of Chemical Engineering* 29(3): 471–481.

Dublin (2019) Globe Newswire – The 'Industrial Enzymes Market – Growth, Trends, and Forecast (2019 – 2024)'. https://www.researchandmarkets.com

Fernandes P (2010) Enzymes in food processing: A condensed overview on strategies for better biocatalysis. *Enzyme Research* doi:10.4061/2010/862537.

Fischetti V, et al. (2002) Use of bacterial phage associated lysing enzymes for treating bacterial infections of the mouth and teeth. United States Patent. Patent No.: US 6,335,012 B1. Accessed 1 Jan 2002.

Garg G, Singh A, Kaur A et al (2016) Microbial pectinases: an ecofriendly tool of nature for industries. *3 Biotech* 6(1): 47–59.

Gayathri VR, Perumal P, Mathew LP, Prakash B (2013) Screening and molecular characterization of extracellular lipase producing Bacillus species from coconut oil mill soil. *International Journal of Science and Technology* 2(7):502–509.

Ghaffarinia A, Jalili C, Riazi-Rad F, et al. (2014) Anti-inflammatory effect of chymotrypsin to autoimmune response against CNS is dose-dependent. *Cellular Immunology* 292:102–108.

Gurung N, Ray S, Bose S, Rai V (2013) A broader view: Microbial enzymes and their relevance in industries, medicine and beyond. *BioMed Research International*: 1–18. 10.1155/2013/329121.

Hajji M, et al. (2008) Optimization of alkaline protease production by Aspergillus clavatus ES1 in Mirabilis jalapa tuber powder using statistical experimental design. *Applied Microbiology and Biotechnology* 79: 915–923.

Hasan F, Shah AA, Javed S, Hameed A (2010) Enzymes used in detergents: Lipases. *African Journal of Biotechnology* 9(31): 4836–4844. doi:10.5897/AJBx09.026.

Hemachander C, Puvanakrishnan R (2000) Lipase from Ralstonia pickettii as an additive in laundry detergent formulations. *Process Biochemistry* 35: 809–814.

Joseph B, Ramteke PW, Thomas G, Shrivastava N (2007) Standard Review Cold-active microbial Lipases: a versatile tool for industrial applications Biotechnol. *Molecular Biology Reviews* 2: 39–48.

Kamini NR, Hemchander C, Geraldine J, et al (1999) Microbial enzyme technology as an alternative to conventional chemical in leather industry. *Current Science* 76: 101.

Kashyap DR, Vohra PK, Chopra S (2001) Applications of pectinases in the commercial sector: A review. *Bioresource Technology* 77: 215–227.

Kembhavi AA, Kulkarni A, Pant A (1993) Salt-tolerant and thermostable alkaline protease from Bacillus subtilis NCIM no 64. *Applied Biochemistry and Biotechnology* 38:10–22.

Kim MH, Kim HK, Lee JK, Park SY, Oh TK (2000) Thermostable lipase of Bacillus stearothermophilus: high-level production, purification and calcium-dependent thermo stability. *Bioscience, Biotechnology, and Biochemistry* 64: 280–286.

Kobayashi S (2010) Lipase-catalyzed polyester synthesis-A green polymer chemistry. *Proceedings of the Japan Academy. Series B, Physical and Biological Sciences* 86(4): 338–365.

Kumar S (2015) Role of enzymes in fruit juice processing and its quality enhancement. *Advances in Applied Science Research* 6(6): 114–124.

Le Roes-Hill M, Prins A (2016) *Biotechnological potential of oxidative enzymes from Actinobacteria.* doi:10.5772/61321.

Lei XG, Stahl CH (2001) Biotechnological development of effective phytases for mineral nutrition and environmental protection. *Applied Microbiology and Biotechnology* 257:474–481.

Lincoln L, More SS (2017) Screening and enhanced production of neutral invertase from Aspergillus sp. by utilization of molasses – A by-product of sugarcane industry. *Advances in Biology Research* 8: 103–110.

Meghwanshi GK, Vashishtha A (2018) Biotechnology of Fungal Lipases. In *Fungi and their Role in Sustainable Development: Current Perspectives*, pp. 383–411. Springer Nature Singapore Pte Ltd. ISBN: 9789811303937.

Niyonzima FN, More SS (2014a) Concomitant production of detergent compatible enzymes by Bacillus flexus XJU-1. *Brazilian Journal of Microbiology* 45: 903–910.

Niyonzima FN, More SS (2014b) Concomitant production of detergent compatible enzymes by Bacillus flexus XJU-1. *Brazilian Journal of Microbiology* 45(3):903–910.

Niyonzima FN, More SS (2019) Coproduction of detergent compatible bacterial enzymes and stain removal evaluation. *Journal of Basic Microbiology* 55 (2015): 1–10. Citation: Francois Niyongabo Niyonzima. Production of Microbial Industrial Enzymes. *Acta Scientific Microbiology* 2(12): 75–89.

Pachauri P, et al. (2018) Kinetic study and characterization of cellulase enzyme from isolated Aspergillus niger subsp. awamori for cellulosic biofuels. *Journal of Science and Industrial Research* 77: 55–60.

Patel AK, Singhania RR, Pandey A, Eds. (2017) *Biotechnology of Microbial Enzymes: Production, Biocatalysis and Industrial Applications*, pp. 13–41. Academic Press Books.

Prasanna HN, et al. (2016) Optimization of cellulase production by Penicillium sp. *3 Biotech* 6: 162–172.

Riffel A (2007) Production of an extracellular keratinase from Chryseo bacterium sp. growing on raw feathers. *Electronic Journal of Biotechnology* 8:35–42.

Robles-Medina A, González-Moreno PA, Esteban-Cerdán L, Molina-Grima E (2009) Biocatalysis: towards ever greener biodiesel production. *Biotechnology Advances* 27(4): 398–408.

Saibabu V, et al. (2013) Isolation, partial purification and characterization of keratinase from Bacillus megaterium. *International Research Journal of Biological Sciences* 2: 13–20.

Salihu A, Alam MZ, AbdulKarim MI, Salleh HM (2012) Lipase production: an insight in the utilization of renewable agricultural residues. *Resources, Conservation and Recycling* 58: 36–44.

Saran S, Mahajan RV, Kaushik R, Isar J, Saxena RK (2013) Enzyme mediated beam house operations of leather industry: a needed step towards greener technology. *Journal of Cleaner Production* 54:315–322.

Saxena RK, Ghosh PK, Gupta R, Davidson W S, Bradoo S, Gulati R (1999) Microbial lipases: potential biocatalysts for the future industry. *Current Science* 77: 101–115.

Sharma CK, Sharma PK, Kanwar SS (2012) Optimization of production conditions of lipase from B. licheniformis MTCC-10498. *Research Journal of Recent Sciences* 1(7): 25–32.

Sharma D, Kumbhar BK, Verma AK, Tewari L (2014) Optimization of critical growth parameters for enhancing extracellular lipase production by alkalophilic Bacillus sp. *Biocatalysis and Agricultural Biotechnology* 3(4): 205–211.

Silva WOB, Mitidieri S, Schrank A, Vainstein MH (2005) Production and extraction of an extracellular lipase from the entomopathogenic fungus Metarhizium anisopliae. *Process Biochemistry* 40: 321–326.

Song J, Tao W, Chen W (2003) Kinetics of enzymatic unhairing by protease in leather industry. *Journal of Cleaner Production* 19:325–331. Alessandro Riffel, Adriano Brandelli. Dehairing activity of extracellular proteases produced by keratinolytic bacteria. *J. of Chem. Technol. & Biotech.*, 78(8):855–859.

Souza PNC, et al. (2015) Optimization of culture conditions for tannase production by Aspergillus sp. gm4 in solid state fermentation. *Acta Science* 37: 23–30.

Sumi H et al (1987) A novel fibrinolytic enzyme (nattokinase) in the vegetable cheese Natto; a typical and popular soybean food inthe Japanese diet. *Experientia* 43: 1110–1111.

Usha KY, et al. (2014) Enhanced production of ligninolytic enzymes by a mushroom Stereum ostrea. *Biotechnology Research International* 2014: 815495.

Veerapagu M, Narayanan AS, Ponmurgan K, Jeya KR (2013) Screening, selection, identification, production, and optimization of bacterial lipase from oil spilled soil. *Asian Journal of Pharmaceutical and Clinical Research* 6(supplement 3):62–67.

Vellard M (2003) The enzyme as drug: application of enzymes as pharmaceuticals. *Current Opinion in Biotechnology* 14: 444–450.

Wang H, et al. (2012) Screening and characterization of a novel alkaline lipase from Acinetobacter calcoaceticus 1–7 isolated from Bohai bay in China for detergent formulation. *Brazilian Journal of Microbiology* 43: 148–156.

Xiao A, et al. (2015) Statistical optimization for tannase production by Aspergillus tubingensis, in solid-state fermentation using tea stalks. *Electronic Journal of Biotechnology* 18: 143–147.

Zaidi KU, Ali AS, Ali SA et al (2014) Microbial tyrosinases: promising enzymes for pharmaceutical, food bioprocessing, and environmental industry. *Biochemistry Research International* doi:10.1155/2014/854687

Zhang H, Sang Q, Zhang W (2012) Statistical optimization of chitosanase production by Aspergillus sp. QD-2 in submerged fermentation. *Annales de Microbiologie* 62(1): 193–201.

2

Applications of Industrially Important Enzymes

Monalisa Padhan
Siddheswar College, Amarda Road, Balasore, Odisha, India

Sushri Priyadarshini Panda
Sambalpur University, Jyoti Vihar, Burla, Odisha, India

CONTENTS

2.1 Introduction: Enzymes as Industrial Biocatalysts

Enzymes are proteins that catalyse chemical reactions and are commonly utilised to speed up industrial processes and product manufacturing, and these enzymes are recognized as industrial enzymes (Zhu et al., 2011). Over the

DOI: 10.1201/9781003292333-2

last few decades, enzymes have become an attractive choice for conventional catalysts in numerous industrial processes. There are many industries in which enzymes are used, such as detergents, animal nutrition, food processing (Kumar et al., 2012), juice and flavour (Gupta et al., 2014; Verma et al., 2013), medication, pharmaceuticals, cosmetics, silk, leather, chemical, and research and development (Kumar and Singh, 2012). Industrial enzymes can be categorized into three types based on the purpose of use: food enzymes, feed enzymes and technical enzymes. The classification of enzymes is described in Figure 2.1. Technical enzymes are utilized in detergents, textiles, fuel, pharmaceuticals, and a variety of small industries. Among the many industries that use food enzymes are the brewing, baking soda, beverage, alcohol, dairy, and oils and fats industries (Kumar et al., 2014a).

As per a report released by the Austrian Federal Environment Agency (AFEA, 2002), there are about 158 enzymes utilized in food industries, 64 enzymes used in technical industries, and 57 enzymes used in the animal feed industry, of which 24 enzymes are used in all three sectors (Kumar et al., 2014b). A brief overview of enzyme applications in these three sectors of the global enzyme industry will be provided in this chapter.

Human civilization has utilized microbes since ancient times, for example, by the Babylonians and Sumerians for the production of alcohol from

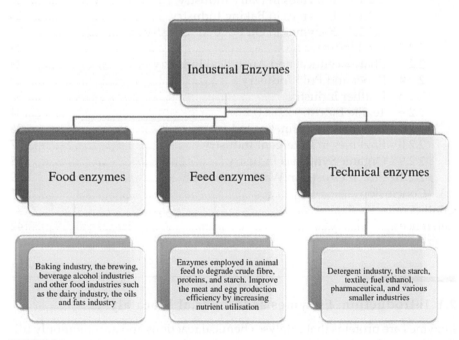

FIGURE 2.1
Classification of industrially important enzymes.

barley as early as 6,000 BCE using yeast (Singh et al., 2016). Gene manipulation has enabled the cultivation of enzymes of microbial origin, particularly those suited to industrial applications. Microbial enzymes are found in many industries, including textiles, pharmaceuticals, paper, leather, and a variety of other fields. There is a rapid rise in the popularity of this method over conventional methods because it is more economic, less harmful to the environment, and generates products with better quality (Singh et al., 2016, Kamini et al., 1999, Gurung et al., 2013). An increased demand for consumer goods, the need to reduce costs, the depletion of natural resources, and environmental safety encourage the application of microbial enzymes in industrial sectors (Choi et al., 2015; Singh et al., 2016).

Carbohydrases, proteases, and lipases are the three main categories of industrial enzymes. Amylases, along with cellulases, glucose isomerase, glucose oxidase, pectinases, xylanases, invertase, galactosidase, and other carbohydrases, make up the carbohydrase group (de Castro et al., 2018). Proteases, amylases, and lipases, which are all hydrolytic enzymes, are the most commonly used industrial enzymes. The detergent, starch processing, animal feed, leather, and food industries all use industrial enzymes. Proteases are the commonly applied enzymes in the detergent and dairy industries, among others. Amylases, cellulases, and other carbohydrases are used in a variety of industries, including starch, textiles, detergents, and baking (Godfrey and West, 1996). Because of their use in cotton processing, detergent enzymes, paper recycling, juice extraction, and animal feed additives, cellulases are currently the third most valuable industrial enzyme in terms of commercial value (Wilson, 2009). Few enzymes have found widespread applications (Figure 2.2), ranging from biobleaching of paper to efficient oil and gas recovery (Bansal et al., 2014).

α-amylase and glucoamylase are the two amylolytic enzymes with the most representative markets. *Bacillus licheniformis* is a major commercial source of

FIGURE 2.2
Application of industrial enzymes in different industries.

α-amylase, while *Aspergillus oryzae* and *Aspergillus niger* are two of the most prevalent glucoamylase hosts (de Castro et al., 2018).

Industrial enzymes are now being used commercially combined with new applications, resulting in significant resource savings, such as in raw material and water consumption, as well as an increase in energy efficiency, which benefits both the industry and the environment. This will continue to be critical in preserving and improving the quality of life we have today, as well as ensuring the sustainability of future generations (Zhu et al., 2011).

2.2 The Use of Enzymes in Industrial Processes

2.2.1 Food Industry

It's been a long tradition to use enzymes or bacteria in food preparation. Novel enzymes with a wide variety of uses and specificity have been developed as technology has progressed. Since ancient times, microorganisms have been utilised in food fermentation, and today, fermentation technologies are employed in the production of a broad range of foods (Soccol et al., 2005). Microbial enzymes are used in the food industry because they are more stable than animal and plant enzymes (Raveendran et al., 2018).

The first applications date back to at least 6,000 BCE, with beer brewing, bread baking, and cheese and wine manufacture, while the first planned microbial oxidation goes back to 2,000 BCE, with the production of vinegar (Chaudhary et al., 2015; Poulsen and Klaus Buchholz, 2003; Vasic-Racki, 2006; Schafer et al., 2002).

Enzymes were first extracted from live cells in the twentieth century, resulting in large-scale commercialization and broader appliance in the food business. Today, the most major source of commercial enzymes is microorganisms. Because of their capacity to function as catalysts, enzymes have always been crucial in food technology, changing basic ingredients into better food products. Food processing enzymes are employed in the production of pre-digested meals, starch processing, meat processing, the dairy sector, and wine production (Chaudhary et al., 2015).

2.2.1.1 Enzymes in Dairy Industry

India is the world's largest milk producer, and the abundance of milk in the nation has prompted the food and dairy industries to use biochemical and enzymatic procedures to turn liquid milk into value-added goods. Rennet was the first exogenous enzyme to be used in the processing of food and it was used to make cheese. Proteinases have found new uses in dairy technology in recent years, such as accelerating the ripening of cheese, modifying functional characteristics, and preparing dietic products (IDF, 1990).

Because aminopeptidases may release single amino acid residues from oligopeptides generated by extracellular proteinase activity, they are vital for the production of taste in fermented milk products (Law and Haandrikman, 1997). In the dairy food sector, lipases and proteases play an important role. Catalase, glucose oxidase, superoxide dismutase, sulphydryl oxidase, lysozymes and lactoperoxidase are other small enzymes with limited uses in dairy processing. For food preservation, glucose oxidase and catalase are frequently combined (Chaudhary et al., 2015).

Microbial enzymes have been used in the dairy sector to produce a variety of products, including yoghurt, cheese, syrup, and bread (Abada, 2019: 62). Animal rennet, plant-derived coagulants, microbial coagulants, and genetically modified chymosin are the four primary types of milk coagulants. The two primary enzymes, chymosin and pepsin, are found in animal rennet in a ratio that varies depending on the age of the animals when slaughtered (Trani et al., 2017: 166). Rennet is a well-known exogenous enzyme that has been utilised in dairy processing since 6,000 BCE. Rennin causes the milk to coagulate by both enzymatic and non-enzymatic activity (Ozatay, 2020).

Different animal or microbe-derived lipases produced clear cheese with low bitterness, better taste, and robust malodours, whereas proteinases in combination with lipases or/and peptidases produced cheeses with good flavour and low bitterness. Lipases from *Mucor miehei* or *Aspergillus niger* are employed in Italian cheeses to give them a richer taste by increasing the quantity of free butyric acid in the milk before adding the rennet (Patel et al., 2016: 67).

The capacity of transglutaminase (TG) to enhance the structure of protein gels has lately piqued the curiosity of food scientists. TG catalyses the posttranslational modification of proteins by forming covalent cross-links between glutamine and lysine residues via transamidation of accessible glutamine residues. The addition of transglutaminase to milk causes casein and whey proteins to crosslink, improving the strength of milk gels (Trani et al., 2017: 170).

2.2.1.2 Enzymes in Baking Industry

For hundreds of years, yeasts and enzymes have been utilized in baking factories to produce a wide range of high-quality goods (Hamer, 1995). Enzymes can be added singly or in diverse mixes at a very low level to produce baked foods that operate synergistically.

Baking involves the usage of enzymes from three different sources: endogenous enzymes found in wheat, enzymes linked to dominating microbe metabolic activity, and exogenous enzymes introduced to the dough (Di Cagno et al., 2003).

Lipolytic enzymes are becoming increasingly used in the baking industry. Recent discoveries show that phospholipases can be utilised to enhance or replace standard emulsifiers since they breakdown polar wheat lipids to create emulsifying lipids (Meshram et al., 2019). In the baking industry, malt is one of the most common sources of enzymes. It contains a number of enzymes, like diastase, that can be put to use as a substitute for low natural a-amylase levels (Hamer, 1995).

A minimal quantity of amylase activity is required in wheat flour. Exogenous amylases can be utilized in wheat to compensate for a lack of amylase activity.

2.2.1.3 Enzymes in Other Food Industry

Ingredient production and texture modification are common uses of enzymes in the food business, with commercial applications including high-fructose corn syrup production, low-lactose milk production, beverage clarifying, baking, brewing, and meat tenderization. Various biopolymers are degraded using a variety of food enzymes. They are preferred over competing chemical therapies because of their selectivity and high response rates under moderate reaction circumstances. Hydrolases, oxidoreductases, and isomerases are the three types of industrial food enzymes. Proteases, amylase, glucoamylase, pectinases, and cellulases (all hydrolases) are bulk enzymes generated mostly by *Bacillus* and *Aspergillus* microbial species. After removing the cells towards the end of the production phase, extracellular enzymes produced by these microorganisms may be readily extracted from squandered culture broth (James et al., 1996; Kula, 1987).

The increase in soluble nitrogen molecules essential for the brewing method is caused by proteinases and exopeptidases found in malt. Exogenous protease preparations must be employed to replace the lost proteolytic activity if unmalted grain is utilised. The use of microbial enzymes in the brewing business is somewhat limited; however, the enzyme acetolactate decarboxylase, which has been extracted from numerous microorganisms is used to catalyse the breakdown of oc-acetolactic acid to acetoin. The oxidative transformation of a-acetolactic acid into diacetyl and the subsequent reduction of diacetyl to acetoin by yeast reductases are generally rate-limiting for the overall maturation process during beer maturation (James et al., 1996). The applications of enzyme in different food industries are detailed in Table 2.1.

2.2.2 Feed Industry

In 1925, the first mention of enzymes in animal feed was made (Hervey 1925; Bedford, 2018). The first step involved employing enzymes to improve nutrient digestibility, with the goal of eliminating anti-nutritive non-starch

TABLE 2.1

Applications of Enzymes in the Food Industry

Type of Food Industry	Enzyme	Source	Function	Reference
Baking Industry	α-Amylase	*Aspergillus* sp., *Bacillus* sp., *Microbacterium imperiale*	Starch modification, flour adjustment, softness of bread, enhance shelf life of breads, ensuring uniform yeast fermentation	Chaudhary et al. (2015); Singh et al. (2016)
	Protease	*Aspergillus niger*	Hydrolysis of casein, gluten, and different plant and animal proteins, reducing the protein in flour	Raveendran et al. (2018); Chaudhary et al. (2015)
	Xylanase	*Aspergillus niger*	Conditioning of the dough	Singh et al. (2016)
	Lipase	*Aspergillus niger*	Flavour enhancement in cheese products, dough conditioning using in-situ emulsification	Chandrasekaran et al. (2015); Singh and Kumar (2019); Bansal et al. (2014)
	Glucose oxidase	*Penicillium chrysogenum*, *Aspergillus niger*	Dough strengthening	Singh et al. (2016)
	Pentosanase	*Trichoderma reesei*, *Humicola insolens*	Flour treatment for noodles, improvement of bread dough	Chaudhary et al. (2015)
	Oxidoreductase	—	Increased gluten strength, enhancing the dough's gas cell stability	Chaudhary et al. (2015)
	Lipoxygenase	—	Dough strengthening, bread whitening, aroma generation in food industry	Raveendran et al. (2018); Chaudhary et al. (2015); Koeduka et al. (2007)

(Continued)

TABLE 2.1 (CONTINUED)

Type of Food Industry	Enzyme	Source	Function	Reference
Dairy Industry	Neutral proteinase	*Aspergillus oryzae, Bacillus subtilis*	Quicker ripening of cheese	Mehta and Sehgal (2019); Singh et al. (2019)
	Acid proteinase	*Aspergillus sp.*	Coagulation of milk	Singh et al. (2016)
	Lipase	*Aspergillus oryzae, Aspergillus niger*	Quicker ripening of cheese, cheese with a unique flavour	Abada (2019)
	Aminopeptidase	*Lactobacillus sp.*	Quicker ripening of cheese	Singh et al. (2016)
	Transglutaminase	*Streptoverticillium sp.*	Modify visco-elastic properties	Bansal et al. (2014); Yokoyama et al. (2000)
	Catalase	*Aspergillus niger*	Cheese processing	Singh et al. (2016)
	Bovine chymosin	Abomasum	Cheese manufacturing	Bhoopathy (1994); Chaudhary et al. (2015)
	β-galactosidase	*Aspergillus spp., Kluyvennnvces spp.*	To avoid lactose intolerance, lactose broken down to glucose and galactose in milk processing	Chaudhary et al. (2015)
	Proteases (papain, chymosin)	*Aspergillus avamori, Kluyvennmyces lactis*	enhancing the taste of milk and cheese, milk clotting, protein hydrolysis	Chaudhary et al. (2015)
Beverage Industry	Pectinase	*Penicillium funiculosum, Aspergillus oryzae*	Depectinization	Ghosh et al. (2016); Singh et al. (2016)
	Glucose oxidase	*Aspergillus sp.*	Dough strengthening, removal of oxygen from beer	Patel et al. (2017)
	Cellulase	*Aspergillus niger*	Liquefaction of fruits	Sharma et al. (2014)
	Amylase	*Bacillus sp., Aspergillus sp., Streptomyces sp.,*	Hydrolysis of starch	Asoodeh et al. (2010); Singh et al. (2016)
	Protease	*Aspergillus niger*	Limit the production of haze	Nair and Jayachandran (2019)

Industry	Enzyme	Source	Application	References
Brewing Industry	β-Glucanase	*Aspergillus* spp., *Bacillus subtilis*	Limit the production of haze	van Oort (2010); Lalor and Goode (2010)
	Pullulanase	*Klebsiella* sp. and *Bacillus* sp.	Saccharification of starch	Xia et al. (2021); Nisha and Satyanarayana (2016)
	Naringinase	*Aspergillus niger*	Debittering	Gupta et al. (2021); Ni et al. (2012)
	Amyloglucosidase	*Aspergillus niger, Rhizopus* spp.	Increasing glucose content	Chaudhary et al. (2015)
	Protease	*Aspergillus niger*	Malt improvement and also improving yeast growth	Aruna et al. (2014); Raveendran et al. (2018)
	Pentosanase, Xylanase	*Humicola insolens, Trichoderma reesei*	Hydrolysing pentosans of wheat, barley, and malt in addition to beer extraction and filtering	Chaudhary et al. (2015)
Juice Industry	Pectinase	*Aspergillus* spp., *Penicillium funiculosum*	It degrades the cell walls of fruits and is also used to clarify juice from fruits and vegetables	Chaudhary et al. (2015)
	Cellulase and Hemicellulase	*Aspergillus* spp., *Bacillus subtilis, Trichoderma reesei*	It is used to degrade the cell walls of fruits and vegetables, as well as to reduce viscosity	Chaudhary et al. (2015)
	Laccase	*Trichoderma* sp., *Bacillus licheniformis, Funalia trogii*	Clarification of juices, flavour enhancer (beer); increase of browning susceptibility during storage	Gochev and Krastanov (2007); Raveendran et al. (2018); Sadhasivam et al. (2008); Li et al. (2017); Tonin et al. (2016)
	α-Amylase and glucoamylase	*Aspergillus niger, Aspergillus awamori, Rhizopus oryzae*	It catalyses the hydrolysis of alpha 1,4 glycosidic bonds in starch polysaccharides to produce dextrins, oligosaccharides, maltose, and D-glucose, resulting in high yields	Raveendran et al. (2018); Coutinho and Reilly (1997)

(Continued)

TABLE 2.1 (CONTINUED)

Type of Food Industry	Enzyme	Source	Function	Reference
	Naringinase and limoninase	Aspergillus sp., Circinella, Eurotium, Fusarium, Penicillium, Rhizopus, Trichoderma, Bacillus sp., Burkholderia cenocepacia, Bacteriodes distasonis, Thermomicrobium roseum, Pseudomonas paucimobilis	Acting on the chemicals in citrus juices that induce bitterness	Raveendran et al. (2018); Puri (2012); Singh et al. (2016)
Starch and Sugar Industry	α-Amylase	Aspergillus sp., Bacillus sp.	Starch liquefaction and saccharification	Raveendran et al. (2018); Singh et al. (2016)
	Xylanase	Aspergillus niger, Streptomyces sp., Bacillus sp., Pseudomonas sp., Fusarium sp., Penicillium sp.	In cereal grains and lignocellulosic biomass, it is necessary for the hydrolysis of xylan polymers	Raveendran et al. (2018); Singh et al. (2016); Sanghi et al. (2010); Bajaj and Singh (2010); Sharma and Chand (2012); Nair et al. (2008)
	Pullulanase	Klebsiella sp., Bacillus sp.	Saccharification	Singh et al. (2016)
	Invertase	Saccharomyces cerevisiae or S. carlsbergensis	Production of invert sugar syrup, sucrose hydrolysis	Bansal et al. (2014)
	α-Acetolactate decarboxylase	Bacillus subtilis	Converting acetolactate to acetoin reduces the time it takes for wine to mature	Chaudhary et al. (2015)

polysaccharides like arabinoxylans and β-glucans from broiler diets made up of viscous grains like barley, wheat, rye, or triticale (Choct, 2006).

Enzymes are employed in animal feed to degrade crude fibre, proteins, starch, and phytates, and because they are proteins, they are digested or ejected by the animal, leaving no trace on meat or eggs. Feed enzymes improve meat and egg production efficiency by increasing nutrient utilisation and lowering animal excreta/waste (Ojha et al., 2019). Several studies have shown that adding degradative enzymes like amylases and proteases to animal feed can boost nutritional availability (Walsh et al., 1993; Chesson, 1987; Lyons, 1991). The addition of certain enzymes to the diet may help to reduce the overall polluting effect of animal excreta. This is especially true in the case of dietary phosphorus, as monogastrics are unable to absorb a considerable fraction of it (Walsh et al., 1993). The use of glycanases in birds fed viscous grains has several advantages: (a) a reduction in manure output containing a high amount of undigested nutrients and (b) relief of problems related to excretion, such as increased percentage of dirty eggs, increased gas production, and increased fly and rodent populations in the shed (Choct, 2006). The enzymes used in the feed industry are given in Table 2.2.

2.2.3 Pharmaceutical and Analytical Industry

Enzymes have a range of significant and critical functions in the pharmaceutical and diagnostic industries. These enzymes are frequently applied as therapeutic drugs for enzymatic deficiency and gastrointestinal illnesses, as well as in diagnostic processes like diabetes testing kits and ELISA (enzyme-linked immunosorbent assay) (Singh et al., 2016; Mane and Tale, 2015).

Tyrosinases' monophenol hydroxylase and diphenoloxidase activities provide the foundation for a broad range of biotechnological appliances in pharmaceutical industries for the production of o-diphenols such as L-DOPA (dopamine) which is used in the treatment of patients suffering from Parkinson's disease. In melanoma patients, it has also been explored as a marker (Gradilone et al., 2010). Microbial tyrosinases are useful tools for today's pharmaceutical technology because of all of these characteristics. The primary enzyme in melanin synthesis is tyrosinase. Radiation protection, cation exchangers, drug transporters, antiviral agents, antioxidants, and immunogens are all examples of synthetic melanin's use (Zaidi et al., 2014). Different enzymes used in the treatment of disorders are listed in Table 2.3.

2.2.4 Paper and Pulp Industry

Since the mid-1980s, the usage of enzymes in the pulp and paper sector has increased dramatically. The prebleaching of kraft pulp is now the most important use of enzymes. The most efficient enzymes for this function have been discovered to be xylanase enzymes. Several mills across the world are

TABLE 2.2

Enzymes Used in Animal Feed Industry

Enzymes in Use	Applications	Microorganisms	Reference
Phytase	Release of phosphorous by hydrolysing phytic acid	*Aspergillus niger*	Singh et al. (2016)
β-Glucanase	Digestive aid	*Aspergillus niger*	Singh et al. (2016)
Cellulase	Breaking down of cellulose/hemicellulose matter present in the animal feed	*Aspergillus niger, Trichoderma atroviride*	Singh et al. (2016)
Xylanase	Improved starch digestibility	*Bacillus* sp., *Aspergillus* sp., *Bacillus* and *Paenibacillus* sp.	Raveendran et al. (2018); Chaudhary et al. (2015); Singh et al. (2016); Sukumaran et al. (2005)
α-Amylase	Aiding in digestion of starch present in animal feed	*Aspergillus* spp., *Bacillus* spp. *Microbacterium irnperiale*	Pandey et al. (2000); Chaudhary et al. (2015)
Protease	It is designed to break down proteins present in the animal feed	*Bacillus subtilis, Aspergillus niger,* Rhizopus sp.	Chaudhary et al. (2015); Singh et al. (2016)
Lipase	It is designed to break down fats present in animal feed	*Aspergillus niger, A. oryzae, Rhizomucor miehei, Penicillium roqueforti, Rhizopus* spp., *Bacillus subtilis*	Chaudhary et al. (2015); Singh et al. (2016)
Galactosidase	Viscosity reduction in lupins and grain legumes used in animal feed	*Aspergillus* spp., *Kluyvennnvces* spp.	Chaudhary et al. (2015)

currently using xylanase prebleaching technology. In the early 1990s, the enzymatic pitch control approach utilising lipase was implemented as a normal operation in a large-scale paper-making process, marking the first time an enzyme had been effectively utilised in the real papermaking process. Using enzymes to improve pulp drainage is a common technique at mill scale (Bajpai, 1999).

The paper and pulp industries produce a lot of effluents which pollute both the water and the land. The effluent has a substantial impact on the environment because it comprises harmful organo-chlorinated chemicals produced throughout various phases of the papermaking process. Biotechnology

TABLE 2.3

Applications of Industrial Enzymes as Therapeutics

Disorders	Enzymes Used	Sources	References
Anti-inflammatory	Serrapeptase, superoxide dismutase	*Lactobacillus* sp., *Corynebacterium glutamicum, Nocardia* sp.	Sabu (2003); Singh et al. (2016)
Collagenase	Skin ulcers	*Clostridium* sp.	Sabu (2003)
Antitumor	L-glutaminase, L-arginase, L-asparaginase, galactosidase	*E. coli, Acinetobacter* sp., *Beauveria bassiana, Pseudomonas* sp.	Sabu (2003); Singh et al. (2016)
Anticoagulants	Urokinase, streptokinase	*Bacillus subtilis, Streptococci* sp.	Sabu (2003); Singh et al. (2016)
Synthesis of antibiotics	penicillin oxidase	*Penicillium* sp.	Sabu (2003); Singh et al. (2016)
Detoxification	Rhodanese, laccase	*Pseudomonas aeruginosa, Trametes versicolor*	Sabu (2003); Singh et al. (2016)
Antioxidants	Catalase, superoxide dismutases, glutathione peroxidases	*Corynebacterium glutamicum, Lactobacillus* sp.	Sabu (2003); Singh et al. (2016)
Antibiotic resistance	β-Lactamase	*Citrobacter* sp., *Klebsiella pneumoniae*	Sabu (2003); Singh et al. (2016)
Digestive disorders	Lipase, α-amylase	*Aspergillus oryzae, Candida lipolytica, Bacillus* spp.	Sabu (2003); Singh et al. (2016)
Antiviral	Serrapeptase, ribonuclease	*Saccharomyces cerevisiae*	Sabu (2003); Singh et al. (2016)
Cyanide poisoning	Rhodanase	*Sulfobacillus sibiricus*	Singh et al. (2016)

provides several cost-effective and pollution-free options, such as using an enzyme cocktail to significantly reduce waste-laden water resources. Laccase, lignin peroxidase, manganese peroxidase, xylanase, lipase, cellulose, protease, and amylase are enzyme cocktails that have promising potential as biobleaching agents since they are very efficient, biorenewable, gentle, non-polluting, selective, and affordable (Kumar et al., 2022).

Because of the environmental risks and illnesses induced by the release of adsorbable organic halogens, cellulase-free xylanases are significant in pulp biobleaching as substitutes for harmful chlorinated chemicals (Walia et al., 2017). The most significant enzymes that may be employed in the pulp and paper processes are cellulase, xylanase, laccase, and lipase (Table 2.4). Further, the research on enzymes has focused on improving bleaching with

TABLE 2.4

Applications of Industrial Enzymes in the Pulp and Paper Industry

Enzymes in Use	Applications	Microorganisms	Reference
Cellulase	Benefits include improved pulp cleanliness, Improvements in drainage	*Aspergillus niger,* *Bacillus* sp.	Kuhad et al. (2011); Bajpai (2012)
Xylanase	Kraft pulp bleaching, Increasing the effectiveness of bleach	*Clostridium thermocellum, Trichoderma reesei, Aureobasidium pullulans, Thermomyces lanuginosus,*	Chapla et al. (2012); Bansal et al. (2014)
α-Amylase	Enzymes are used especially in modification of starch. Starch improves the strength, stiffness and erasability of paper. Deinking, drainage improvement	*Bacillus licheniformis*	Singh et al. (2016)
Lipase	Pitch control	*Candida Antarctica, Candida rugosa*	Mehta et al. (2021); Anobom et al. (2014); Casas-Godoy et al. (2012)
Ligninase	Remove lignin to soften paper	*Trametes versicolor, Phlebia radiata*	Kumar and Chandra. (2020)
Laccase	Bleach to improve brightness, non-chlorine bleaching, delignification	*Trametes hirsuta*	Singh et al. (2016)
Mannanases	Degrading the residual glucomannan to increase brightness	*Pseudomonas fluorescens; Clostridium thermocellum*	Clarke et al. (2000)
Protease	Biofilm removal	*Bacillus subtilis*	Danilova and Sharipova (2020)
Cellobiohydrolase	Mechanical pulping	*Clostridium* sp., *Fervidobacterium islandicum; Botrytiscinerea*	Zafar et al. (2021); Chaudhary et al. (2015)
Polygalacturonase	Debarking, mechanical pulping	*Aspergillus niger*	Bajpai (2018)

xylanases and fibre modification with cellulases (Demuner et al., 2011). Biopulping makes use of fungi-produced enzymes to minimise the usage of chemicals in the pulping step of wood chips (Singh and Chen 2008; Ferraz et al. 2008), boost fibre production, reduce subsequent refining energy needs, or give particular fibre changes (Ferraz et al. 2008; Singh and Chen 2008; Torres et al., 2012; Kenealy and Jeffries 2003; Singh et al. 2010).

2.2.5 Leather Industry

Leather manufacturing is another industry that is being hampered by industrial pollution since it uses several chemicals in the pretanning process, such as lime, sodium sulphide, salts, and solvents. The wastes created during dehairing processes account for about a third of the pollution caused by the leather industry. The use of enzymes in different pre-tanning procedures including soaking, dehairing, bating, and degreasing might be a feasible alternative to traditional methods for reducing pollution produced by tannery effluents. Table 2.5 lists the most regularly used enzymes in leather manufacturing (Bansal et al., 2014).

Choudhary et al. (2004) predicted that the leather sector will have a lot of potential for using lipase, protease, amylase, trypsin, pepsin, rennin, and

TABLE 2.5

Applications of Industrial Enzymes in Leather Industry

Enzymes in Use	Applications	Microorganisms	Reference
Neutral protease	Soaking, dehairing	*Aspergillus flavus*, *Aspergillus niger*, *Bacillus subtilis*, and *Bacillus cereus*	Madhavi et al. (2011)
Acid and alkaline protease	Bating, dehairing	*Aspergillus flavus*, *Conidiobolus coronatus*, and *Alcaligenes faecalis*	Souza et al. (2015); Laxman et al. (2005); Malathi and Chakraborty (1991)
Alkaline and acid lipase	Removal of the fat and grease from the interfibrillary spaces of the skins; smoothen the leather surface for the production of leather garments	*Aspergillus flavus*, *Aspergillus oryzae*	Khambhaty (2020)
Amylase	Fibre splitting	*Bacillus* sp., *Aspergillus* sp.	Pandi et al. (2016)

glutaminase throughout processing and effluent treatment up until the year 2000. Gupta and Ramnani (2006) gave an outline of the possible uses of microbial keratinases in the leather industry, stating that the complicated process of keratinolysis must be better understood, and that innovative, robust keratinases must be investigated. The dehairing efficiency of *Bacillus megaterium* protease has also been researched, with the qualities of preventing collagen damage and reducing pollution making this enzyme a potential choice for the tanning process (Rajkumar et al. 2011; Khambhaty, 2020). Enzymes can be used in the soaking, dehairing, bating, dyeing, degreasing, and effluent and solid waste treatment stages of the leather production process (de Souza and Gutterres, 2012) and are described in Table 2.5.

2.2.6 Textile Industry

Among the most quickly increasing fields in industrial enzymology is the usage of enzymes in textile manufacturing. Amylases, catalases, and laccase are enzymes utilized in the textile industry to remove starch, degrade excessive hydrogen peroxide and lignin, and bleach fabrics. The application of enzymes in textile chemical processing is fast gaining global attention due to its non-toxic and environmentally benign properties, which are becoming increasingly relevant as textile manufacturers strive to decrease pollution in their operations. The most modern commercial breakthroughs include the use of cellulases in denim manufacturing during its finishing (Araujo et al., 2008). Further, lactases are also used for textile bleaching. Amylases are used in the textile industry to eliminate starch-based sizing for better and consistent wet processing. Amylase breaks down dietary starch into dextrin and maltose. These enzymes have the benefit of being starch specific, indicating they can remove it without harming the support fabric. Desizing operations can be carried out using an amylase enzyme at low temperatures (30–60°C) and at a pH of 5.5–6.5 (Mojsov, 2011; Cavaco-Paulo and Gübitz, 2003).

Hydrolases and oxidoreductases are mostly used in the textile industry for numerous enzymatic applications. Because of the wide range of catalytic activity accessible, high yields, simplicity of genetic modification, consistent supply owing to lack of seasonal changes, and quick growth of microorganisms in low-cost medium, enzymes secreted by microbes are more often used than plant- or animal-derived enzymes (Madhu and Chakraborty, 2017; Wavhal and Balasubramanya, 2011). Because of the darker shade and drainage of synthetic dyes, the wastewater generated by the textile industry is harmful to the biological world. Lipases, pectinases, catalases, peroxidases, ligninases, collagenases, cellulases, amylases, and proteases are just a few of the fungal enzymes (Gupta et al., 2015) that have shown their utility in the manufacturing and construction sectors (Table 2.6). The ability of fungus to degrade dyes has been proven (Karnwal et al., 2019).

TABLE 2.6

Applications of Industrial Enzymes in the Textile Industry

Enzymes in Use	Applications	Microorganisms	References
α-amylase	Desizing of fabric like cotton	*Bacillus licheniformis, Bacillus* sp.	Colomera and Kuilderd (2015); Karnwal et al. (2019)
Cellulase	Finishing of denim and softening of cotton, biofinishing	*Penicillium funiculosum, Aspergillus niger*	Ali et al. (2012)
Pectate lyase	Bioscouring	*Pseudomonas* sp., *Bacillus* sp.	Singh et al. (2016)
Catalase	Elimination of hydrogen peroxide residues after bleaching of cotton fabrics	*Aspergillus* sp.	Karnwal et al. (2019); Amorim et al. (2002)
Laccase	Fabric dyeing, bleaching without chlorine	*Bacillus subtilis*	Singh et al. (2016)
Lipase	Denim finishing and removal of size lubricant	*Candida Antarctica*	Karnwal et al. (2019); El-Shemy et al. (2016)
Protease	Degumming of silk and removal of wool fibre scales	*Bacillus* sp., *Aspergillus niger*	Karnwal et al. (2019); Periolatto et al. (2011)
Collagenase	Biocatalysts in the exhaustion of dyes	*Clostridium histolyticum*	Karnwal et al. (2019); Kanth et al. (2008)
Ligninase	Finishing of wool	*Phlebia radiate, Trametes versicolor*	Karnwal et al. (2019); Young and Yu (1997)
Cutinase	Synthetic fibre modification and Cotton scouring	*Thermomonospora fusca, Pseudomonas mendocina*	Martinez and Maicas (2021)
Glucose oxidase	Bleaching	*Aspergillus niger, Penicillium chrysogenum*	Chaudhary et al. (2015)
Xylanase	Degradation of lignin	*Streptomyces* sp., *Bacillus* sp. and *Pseudomonas* sp.	Raveendran et al. (2018); Sanghi et al. (2010); Bajaj and Singh (2010); Sharma and Chand (2012)
Pectinase	Scouring cotton, retting bast fibres	*Aspergillus oryzae, Penicillium funiculosum*	Bansal et al. (2014)

(Continued)

TABLE 2.6　(CONTINUED)

Enzymes in Use	Applications	Microorganisms	References
Peroxidase	Used as an enzymatic rinse process after reactive dyeing, the oxidative splitting of hydrolysed reactive dyes on the fibre, and biobleaching of important industrial dyes	*Phanerochaete chrysosporium, Coprinus cinereus*	Karnwal et al. (2019); Osuji et al. (2014)
Esterase	Partial hydrolysis of synthetic fibre surfaces, improving their hydrophilicity and aiding further finishing steps	*Bacillus* sp., *Pseudomonas* sp., *Streptomyces* sp., *Thermoanaerobacterium* sp., *Thermoanaerobacterium* sp.	Panda and Gowrishankar (2005); Karnwal et al. (2019); Araujo et al. (2008)
Nitrilases	In the development of polyacrylonitrile preparation for better colouration in textile processing	*Rhodococcus rhodochrous, Brevibacterium imperial, Corynebacterium nitrilophilus, Arthrobacter* sp.	Matama et al. (2007); Tauber et al. (2000); Karnwal et al. (2019); Robinson and Hook (1964)

2.2.7　Enzymes in Cosmetics Industry

Enzymes have been discovered to have uses in the cosmetics market, such as hair colouring, dental cleaning, skin therapy, and facewash. However, due to their safety features and performance, their applications are limited. As a result, the potential of enzymes in cosmetics preparations has been consistently examined.

The enzymes like oxidases, peroxidases and polyphenol oxidases have found their utility in hair dyeing. However, because of concerns with permanency, anti-greying performance, and colour diversity, enzyme hair colouring treatments cannot compete with traditional dyes. Several firms, including L'Oreal, Lion, Henkel, and others, are always attempting to enhance the effectiveness of enzymes in their hair dye products, as seen by the growing number of patent filings relating to enzymatic hair colouring methods (Bansal et al., 2014).

Enzymes in the skin are responsible for uncoupling complex inactive compounds and converting them into simpler, more active ones. Proteases, for example, decouple or hydrolyse proteins, whereas glycosidases increase ceramide enrichment in the epidermis and tyrosinase aids melanin production. Enzymatic reactions regulate keratinisation processes, verify inter-corneocytar cohesion, stimulate tanning and auto-photoprotection, act on the metabolism of sebaceous glands and adipocytes, whiten age spots,

TABLE 2.7

Applications of Industrial Enzymes in the Cosmetics Industry

Enzymes in Use	Applications	Microorganisms	Reference
Superoxide dismutase	Skin care, scavenging of free radicals	*Corynebacterium* sp.	Peyrat et al. (2019); El Shafey et al. (2010)
Protease	Dead skin removal	*Aspergillus flavus, Aspergillus niger, Bacillus subtilis*	Naveed et al. (2021); Solanki et al. (2021)
Endoglycosidase	Taking care of teeth and gum tissues	*Mucor hiemalis, Rhodococcus* sp.	Arora (2020); Katoh et al. (2016); Ito and Yamagata (1989); Vashist et al. (2019)
Lipase	Skin care	*Aspergillus flavus, Aspergillus oryzae*	Kavitha et al. (2021)
Laccase	Hair colour	*Trametes versicolor, Bacillus* sp.	Kumar et al. (2018)

stimulate the skin's natural defence mechanisms, or safeguard collagen and elastin fibres on the skin and stratum corneum (Charbonelle, 1999). As a result, enzymes appear to be potential natural active ingredients for some personal care products classified as "functional cosmetics," "cosmeceuticals," or "treatment products" (Goldstein, 2002; Ansorge-Schumacher and Thum, 2013). The enzymes that play major roles in cosmetics are detailed in Table 2.7.

2.2.8 Enzymes in Detergent Industry

Enzymes have become more widely used in industrial applications, particularly in detergents, as a result of recent advances in enzyme technology. Enzymes from microbes grown in a variety of climates, from extreme hot to extreme cold, are being investigated as detergent additives for compatibility tests. Cold-active enzymes with high catalytic activity are found in psychrophiles that develop in cold environments, and their durability under harsh temperatures makes them an excellent eco-friendly and cost-effective detergent additive. A sufficient number of publications on cold-active enzymes with high efficiency and extraordinary properties, such as proteases, lipases, amylases, and cellulases, are accessible. These enzymes have become the most popular detergent additives due to their higher thermostability and alkaline stability (Al-Ghanayem and Joseph, 2020).

Lipases are used in domestic and commercial laundry detergents (Kumar et al., 1998), as well as in household dishwashers, where they remove fatty residues and unclog clogged drains (Vulfson, 1994). To boost the efficacy

TABLE 2.8

Applications of Industrial Enzymes in the Detergent Industry

Enzymes in Use	Microorganisms	Applications	References
Alkaline protease	*Paenibacillus tezpurensis, Aspergillus oryzae, Bacillus subtilis, Salinicoccus* sp.	Protein stain removal of blood and grass	Rai et al. (2010); Mokashe et al. (2017)
Alkaline amylase	*Bacillus subtilis, Aspergillus* sp.	Getting rid of carbohydrate stains, sauce, gravy stain	Niyonzima and More (2015)
Alkaline lipase	*Aspergillus oryzae, A. flavus*	Remove stains of fats and oils	Laachari et al. (2015)
Alkaline cellulase	*Aspergillus niger, Bacillus* sp.	Remove particulate soils, soften and improve the colour brightness of the clothes	Kuhad et al. (2011); Bansal et al. (2014)
Cutinase	*Fusarium solani*	Removal of triglyceride	Badenes et al. (2010); Chen et al. (2010)
Mannanase	*Bacillus* sp.	Remove various food stains containing guar gum	Srivastava and Kapoor (2014); Kim et al. (2018)

of detergents, enzymes like proteases, amylases, cellulases, and lipases are added (Hasan et al., 2010; Ito et al., 1998). Enzymes useful in the detergent industry and their sources are described in Table 2.8.

2.2.9 Organic Synthesis Industry

The current state of the art in environmentally friendly organic synthesis utilising enzymes is presented in Green Chemistry. In every case, the enzyme is the crucial component that permits the chemistry and helps to make it green (Itoh and Hanefeld, 2017).

Enzymes are commonly utilised as catalysts in the production of chiral chemical molecules, in particular. Among the asymmetric catalysts, enzymes are frequently the most efficient and eco-friendly. Oxygenases are fascinating enzymes in organic synthesis because they may insert one or two oxygen atoms into organic compounds, resulting in enantio-, chemo-, or regioselective products. Dioxygenases are a diverse set of NAD(P)H-dependent enzymes that are involved in the production of secondary metabolites including flavonoids and alkaloids. They also play a vital role in the natural breakdown

TABLE 2.9

Applications of Industrial Enzymes in the Organic Synthesis Industry

Enzymes in Use	Applications	Microorganisms	References
Lipase	Synthesis of polymers, drugs, biosurfactants and biofuels	*Aspergillus flavus*, *Aspergillus oryzae*	Mehta et al. (2017); Uma and Sivagurunathan (2021)
Glycosyl tranferase	Oligosaccharides synthesis	*Bacillus* sp.	Thombre and Kanekar (2017)
Glucose isomerase	High-fructose corn syrup production	*Streptomyces murinus*, *Corynebacterium* sp.	Desmet and Soetaert (2011)
Acyltransferase	Hydroxamic acids synthesis	*Bacillus* sp.	Pandey et al. (2019); Sogani et al. (2012)
Laccase	Production of textile dyes, flavour agents, cosmetic pigments, and pesticides	*Bacillus subtilis*, *Trametes versicolor*	Dwivedi et al. (2011); Shiroya (2021)

of aromatic molecules (Urlacher and Schmid, 2006). Aminotransferases are significant enzymes that have been employed in the synthesis of amino acids in the past (Strohmeier et al., 2011; Rozell and Bommarius, 2002).

Many distinct reactions are catalysed by human enzymes. The principal use of human enzymes for organic synthesis at the preparative stage is now in the field of drug metabolite production (Winkler et al., 2018). The oxidative conversion of ethanol and retinol to the corresponding aldehydes is the primary function of alcohol dehydrogenases in humans (Winkler et al., 2018; Yin et al., 1999). Monoamine oxidase enzymes (MAOs) in the human body oxidise primary amines such as the neurotransmitters serotonin, dopamine, and epinephrine to imines, which then breakdown into the corresponding aldehydes in the aqueous environment (Weyler et al., 1990). Human variants, like other cytochrome P450 enzymes, have been demonstrated to catalyse epoxidation processes, such as lipid epoxidation (Edin et al., 2015). Enzymes used in the organic synthesis industry are detailed in Table 2.9.

2.2.10 Enzymes used in Waste Treatment

The necessity for alternative waste treatment technologies has been prompted by the introduction of more rigorous criteria for garbage release into the environment. A wide range of enzymes from various plants and microorganisms have been found to play a key role in a variety of waste treatment applications. Enzymes can act on specific recalcitrant contaminants to precipitate or change them into other compounds, allowing them to be removed. They can also alter the features of a waste to make it more treatable or assist in the conversion of trash into value-added goods. Lignin peroxidase, Horseradish peroxidise and a variety of other peroxidases from other sources have all been employed for aqueous aromatic pollutants treatment in the laboratory. Several phenolic compounds have been found to be oxidised by the chloroperoxidase secreted by the fungus *Caldariomyces fumago*. Laccase is synthesized by numerous fungi and appears to be capable of reducing phenolic chemical toxicity via a polymerization process (Karam and Nicell, 1997; Bollag et al., 1988). Nanotubes containing oxidative enzymes such as laccases and peroxidases can also be used to remove recalcitrant contaminants in wastewater (Pandey et al., 2017). Furthermore, because of the wide range of contaminants contained in waste, enzymes with broad substrate specificities will be better suited to waste treatment markets. Organophosphate pesticide hydrolases and phenol-oxidizing enzymes are examples of broad-specificity enzymes that may convert a wide variety of chemicals in a particular class (Aitken, 1993).

According to European Parliament legislation, keratin-rich animal byproducts are categorised as category three waste items. In addition, cattle populations across the world create millions of tonnes of keratinous waste in the form of hide, hair, and hoofs, among other things. Other sources include fur industries, large-scale meat processing plants, and slaughterhouses. Keratinolytic bacteria can hydrolyse keratinous wastes that are "hard to decompose." "Keratinases" is the name given to this novel class of proteases. Keratinases have substituted regular proteases in many industrial applications, such as nematicidal agents, nitrogenous fertiliser manufacture from keratinous waste, animal feed, and biofuel generation, due to their specificity (Verma et al. 2017). The discovery of keratinases has opened a new era in waste management, with industrial applications resulting in green technologies for long-term growth (Pandey et al., 2017). Different enzymes playing different roles in waste treatment with their sources and application are given in Table 2.10.

TABLE 2.10

Applications of Industrial Enzymes in Waste Treatment

Enzymes in Use	Applications	Microorganisms	References
Amidase	Nitriles containing waste degradation	*Rhodococcus erythropolis*	Rathore et al. (2021); Egelkamp et al. (2017)
Amylase	Vegetable waste bioremediation	*Aspergillus* sp., *Bacillus licheniformis*	Janarthanan et al. (2014)
Amyloglucosidase	Hydrolysis of starch for bioremediation	*Aspergillus niger*	Sirisha et al. (2016)

(Continued)

TABLE 2.10   (CONTINUED)

Enzymes in Use	Applications	Microorganisms	References
Nitrile hydratase	Nitriles containing waste degradation	*Rhodococcus* sp.	An et al. (2020); Mukram et al. (2015)
Protease	Keratinic waste bioremediation	*Chrysosporium keratinophilum*	Bohacz (2017); Gopinath et al. (2015)
Manganese peroxidase	Phenolic compounds degradation	*Coprinus cinereus, Phanerochaete chrysosporium*	El-Shora et al. (2017)
Cutinase	Degradation of polycaprolactone and plastics	*Fusarium solani*	Shi et al. (2020)
Lignin peroxidase	Phenolic compounds degradation	*Coprinus cinereus, Phanerochaete chrysosporium*	Pham et al. (2016)
Oxygenase	Halogenated contaminants degradation	*Rhodococcus* sp., *Pseudomonas* sp.	Peng and Shih (2013)

2.3 Conclusion

As previously stated, enzymes are presently employed in a variety of food items and processes, with new applications being introduced on a regular basis. Enzymes are excellent biocatalysts that perform in moderate circumstances, resulting in huge savings in resources like energy and the environment. The evidence clearly reveals that persistent research efforts are being undertaken to improve the effectiveness and diversity of biological agent applications. These efforts have centred on developing new/improved biocatalysts that are more stable, less reliant on metal ions, and less vulnerable to inhibitors and harsh environmental conditions while preserving or generating unique functions.

Biotechnology has developed as a technical revolution throughout the world in the previous two decades as a result of the widening variety of enzyme application industries, which continues to grow. It has had an impact on nearly every aspect of industrial activity, including chemical feedstock, food, feed, the environment, energy, and healthcare. The use of enzymes in practically every industry is expected to rise in the future.

The evidence clearly reveals that persistent research efforts are being undertaken to improve the effectiveness and diversity of biological agent applications. These efforts have been centred on developing new/improved biocatalysts that are more stable, less reliant on metal ions, and less vulnerable to inhibitors and harsh environmental conditions while retaining or evolving the intended activity.

Conflict of Interest

The authors declare that they have no conflict of interest.

References

Abada, E.A. 2019. "Application of microbial enzymes in the dairy industry". In *Enzymes in Food Biotechnology*, 61–72. Academic Press.

AFEA. 2002. *Collection of Information on Enzymes*. Austrian Federal Environment Agency. Available from: Http://Ec.Europa.Eu/Environment/Archives/Dansub/Pdfs/Enzymerepcomplete.Pdf

Aitken, M.D. 1993. "Waste treatment applications of enzymes: Opportunities and obstacles."*The Chemical Engineering Journal* 52(2): 49–58.

Al-Ghanayem, A.A., and Joseph, B. 2020. "Current prospective in using cold-active enzymes as eco-friendly detergent additive." *Applied Microbiology and Biotechnology* 104(7): 2871–2882.

Ali, H., Hashem, M., Shaker, N., Ramadan, M., El-Sadek, B., and Hady, M.A. 2012. "Cellulase enzyme in bio-finishing of cotton-based fabrics: Effects of process parameters." *Research Journal of Textile and Apparel* 16: 57–65.

Amorim, A.M., Gasques, M.D., Andreaus, J., and Scharf, M. 2002. "The application of catalase for the elimination of hydrogen peroxide residues after bleaching of cotton fabrics." *Anais da Academia Brasileira de Ciências* 74: 433–436.

An, X., Cheng, Y., Miao, L., Chen, X., Zang, H., and Li, C. 2020. "Characterization and genome functional analysis of an efficient nitrile-degrading bacterium, *Rhodococcus rhodochrous* BX2, to lay the foundation for potential bioaugmentation for remediation of nitrile-contaminated environments." *Journal of Hazardous Materials* 389: 121906.

Anobom, C.D., Pinheiro, A.S., De-Andrade, R.A., Aguieiras, E.C., Andrade, G.C., Moura, M.V., Almeida, R.V., and Freire, D.M. 2014. "From structure to catalysis: Recent developments in the biotechnological applications of lipases." *BioMed Research International* 2014, 684506.

Ansorge-Schumacher, M.B., and Thum, O. 2013. "Immobilised lipases in the cosmetics industry." *Chemical Society Reviews* 42(15): 6475–6490.

Araujo, R., Casal, M., and Cavaco-Paulo, A. 2008. "Application of enzymes for textile fibres processing." *Biocatalysis and Biotransformation* 26(5): 332–349.

Arora, R. 2020. "Industrial potential of microbial enzymes." In *Microbial Diversity, Interventions and Scope*, edited by Mohit Sharma, Neeta Raj Sharma, and Shiwani Guleria Sharma, 301–318. Springer, Singapore.

Aruna, K., Shah, J., and Birmole, R. 2014. "Production and partial characterization of alkaline protease from *Bacillus tequilensis* strains CSGAB 0139 isolated from spoilt cottage cheese." *International Journal of Applied Biology and Pharmaceutical* 5: 201–221.

Asoodeh, A., Chamani, J., and Lagzian, M. 2010. "A novel thermostable, acidophilic α-amylase from a new thermophilic '*Bacillus* sp. Ferdowsicous isolated from Ferdows hot mineral spring in Iran: Purification and biochemical characterization'." *International Journal of Biological Macromolecules* 46(3): 289–297.

Badenes, S.M., Lemos, F., and Cabral, J.M. 2010. "Assessing the use of cutinase reversed micellar catalytic system for the production of biodiesel from triglycerides." *Journal of Chemical Technology & Biotechnology* 85(7): 993–998.

Bajaj, B.K., and Singh, N.P. 2010. "Production of xylanase from an alkali tolerant *Streptomyces* sp. 7b under solid-state fermentation, its purification, and characterization." *Applied Biochemistry and Biotechnology* 162: 1804–1818.

Bajpai, P. 1999. "Application of enzymes in the pulp and paper industry." *Biotechnology Progress* 15(2): 147–157.

Bajpai, P. 2012. *Biotechnology for Pulp and Paper Processing*,7–13. Springer, New York.

Bajpai, P. 2018. "Biodebarking." In *Biotechnology for Pulp and Paper Processing*, 57–66. Springer, Singapore.

Bansal, S., Gunjan, G., and Ojha, S. 2014. "Industrial enzyme production." *Industrial enzymes trends, scope and relevance*, edited by Vikas Beniwal, Anil Kumar Sharma, 15–32, Nova Science Publishers, New York.

Bedford, M.R. 2018. "The evolution and application of enzymes in the animal feed industry: The role of data interpretation." *British Poultry Science* 59(5): 486–493.

Bhoopathy, R. 1994. "Enzyme technology in food and health industries." *Indian Food Industry* 13: 22–31.

Bohacz, J. 2017. "Biodegradation of feather waste keratin by a keratinolytic soil fungus of the genus Chrysosporium and statistical optimization of feather mass loss." *World Journal of Microbiology and Biotechnology* 33(1): 1–16.

Bollag, J.M., Shuttleworth, K.L., and Anderson, D.H. 1988. "Laccase-mediated detoxication of phenolic compounds." *Applied and Environmental Microbiology* 54.

Casas-Godoy, L., Duquesne, S., Bordes, F., Sandoval, G., and Marty, A. 2012. "Lipases: An overview." *Lipases and Phospholipases*, 3–30.

Cavaco-Paulo, A., and Gübitz, G.M. 2003. *Textile Processing with Enzymes*. Woodhead Publishing Ltd., England.

Chandrasekaran, M., Basheer, S.M., Chellappan, S., Krishna, J.G., and Beena, P.S. 2015. "Enzymes in food and beverage production: An overview." *Enzym Food Beverage Process CRC Press* 25: 133–154.

Chapla, D., Patel, H., Madamwar, D., and Shah, A. 2012. "Assessment of a thermostable xylanase from *Paenibacillus sp.* ASCD2 for application in prebleaching of eucalyptus kraft pulp." *Waste and Biomass Valorization* 3(3): 269–274.

Charbonelle, P. 1999. "Kosmet."80: 32–35.

Chaudhary, S., Sagar, S., Kumar, M., Sengar, R.S., and Tomar, A. 2015. "The use of enzymes in food processing: A review." *South Asian Journal of Food Technology and Environment* 1(3&4): 190–210.

Chen, S., Su, L., Billig, S., Zimmermann, W., Chen, J., and Wu, J. 2010. "Biochemical characterization of the cutinases from *Thermobifida fusca*." *Journal of Molecular Catalysis B: Enzymatic* 63(3–4): 121–127.

Chesson, A. 1987. *Recent Advances in Ammal Nutrition*, edited by P. Garnsworthy, W. Haresign, and D. Cole, 71–89. Butterworth-Heinemann.

Choct, M. 2006. "Enzymes for the feed industry: Past, present and future." *World's Poultry Science Journal* 62(1): 5–16.

Choi, J.M., Han, S.S., and Kim, H.S. 2015. "Industrial applications of enzyme biocatalysis: Current status and future aspects." *Biotechnology Advances* 33(7): 1443–1454.

Choudhary, R.B., Jana, A.K., and Jha, M.K. 2004. "Enzyme technology applications in leather processing." *Indian Journal of Chemical Technology* 11: 659–671.

Clarke, J.H., Davidson, K., Rixon, J.E., Halstead, J.R., Fransen, M.P., Gilbert, H.J. and Hazlewood, G.P. 2000. "A comparison of enzyme-aided bleaching of softwood paper pulp using combinations of xylanase, mannanase and α-galactosidase." *Applied Microbiology and Biotechnology* 53(6): 661–667.

Colomera, A., and Kuilderd, H.2015 "Biotechnological washing of denim jeans." In: Paul, R., (ed), *Denim: Manufacture, Finishing and Applications*, pp. 357–403, Woodhead Publishing.

Coutinho, P.M., and Reilly, P.J. 1997. "Glucoamylase structural, functional and evolutionary relationships." *Proteins* 29(3): 334–347.

Danilova, I., and Sharipova, M. 2020. "The practical potential of Bacilli and their enzymes for industrial production." *Frontiers in Microbiology* 11: 1782.

de Castro, A.M., dos Santos, A.F., Kachrimanidou, V., Koutinas, A.A., and Freire, D.M. 2018. "Solid-state fermentation for the production of proteases and amylases and their application in nutrient medium production." In: Pandey, A., Larroche, C., Soccol, C.R. (eds) *Current Developments in Biotechnology and Bioengineering: Current Advances in Solid-State Fermentation*, Elsevier, pp. 185–210.

De Souza, F.R., and Gutterres, M. 2012. "Application of enzymes in leather processing: A comparison between chemical and coenzymatic processes." *Brazilian Journal of Chemical Engineering* 29: 473–482.

Demuner, B.J., Pereira Junior, N., and Antunes, A. 2011. "Technology prospecting on enzymes for the pulp and paper industry." *Journal of Technology Management & Innovation* 6(3): 148–158.

Desmet, T., and Soetaert, W. 2011. "Enzymatic glycosyl transfer: Mechanisms and applications." *Biocatalysis and Biotransformation* 29(1): 1–18.

Di Cagno, R., De Angelis, M., Corsetti, A., Lavermicocca, P., Arnault, P., Tossut, P., Gallo, G., and Gobbetti, M. 2003. "Interactions between sourdough lactic acid bacteria and exogenous enzymes: Effects on the microbial kinetics of acidification and dough textural properties." *Food Microbiology* 20(1): 67–75.

Dwivedi, U.N., Singh, P., Pandey, V.P., and Kumar, A. 2011. "Structure–function relationship among bacterial, fungal and plant laccases." *Journal of Molecular Catalysis B: Enzymatic* 68(2): 117–128.

Edin, M.L.Cheng, J., Gruzdev, A., Hoopes, S.L., and Zeldin, D.C. 2015. "Cytochrome P450 Struct." In *MechanicalBiochemical*, 4th ed., edited by P.R. Ortiz de Montellano, 881–905. Springer International Publishing, Cham.

Egelkamp, R., Schneider, D., Hertel, R., and Daniel, R. 2017. "Nitrile-degrading bacteria isolated from compost." *Frontiers in Environmental Science* 5: 56.

El Shafey, H.M., Bahashwan, S.A., Alghaithy, A.A., and Ghanem, S. 2010. "Microbial superoxide dismutase enzyme as therapeutic agent and future gene therapy." *Current Research, Technology and Education. Topics in Applied Microbiology and Microbial Biotechnology* 1: 435–443.

El-Shemy, N.S., El-Hawary, N.S., and El-Sayed, H. 2016. "Basic and reactive-dye-able polyester fabrics using lipase enzymes." *Journal of Chemical Engineering & Process Technology* 7: 271.

El-Shora, H.M., Ibrahim, M.E., El-Sharkawy, R.M., and Elmekabaty, M.R. 2017. "Manganese peroxidase from *Trichoderma harzianum* and increasing its efficiency for phenol removal from wastewater." *Journal of Advances in Microbiology* 5: 1–12.

Ferraz, A., Guerra, A., Mendonza, R., Masarin, F., Vicentim, M.P., Aguiar, A., and Pavan, P.C. 2008. "Technological advances and mechanistic basis for fungal biopulping Enzyme." *Microbial Technology* 43: 178–185.

Ghosh, P., Pradhan, R.C., and Mishra, S. 2016. "Optimization of process parameters for enhanced production of Jamun juice using Pectinase (*Aspergillus aculeatus*) enzyme and its characterization." *3 Biotech* 6(2): 1–11.

Gochev, V.K., and Krastanov, A.I. 2007. "Isolation of laccase producing *Trichoderma sp.*" *Bulgarian Journal of Agricultural Science* 13: 171–176.

Godfrey, T., and West, S. 1996. "Introduction to industrial enzymology". In *Industrial Enzymology*, edited by T. Godfrey, and S. West. Cambridge University Press, Cambridge.

Goldstein, M.S. 2002. *Chemistry and Manufacture of Cosmetics*, edited by M.L. Schlossman, 405–415.

Gopinath, S.C., Anbu, P., Lakshmipriya, T., Tang, T.H., Chen, Y., Hashim, U., Ruslinda, A.R., and Arshad, M.K. 2015. "Biotechnological aspects and perspective of microbial keratinase production." *BioMed Research International*. doi: 10.1155/2015/140726

Gradilone, A., Cigna, E., Agliano, A.M., and Frati, L. 2010. "Tyrosinase expression as a molecular marker for investigating the presence of circulating tumor cells in melanoma patients." *Current Cancer Drug Targets* 10(5): 529–538.

Gupta, A., Kumar, V., Dubey, A., and Verma, A.K. 2014. "Kinetic characterization and effect of immobilized thermostable β-glucosidase in alginate gel beads on sugarcane juice." *International Scholarly Research Notices* 2014: 815495.

Gupta, A.K., Yumnam, M., Medhi, M., Koch, P., Chakraborty, S., and Mishra, P. 2021. "Isolation and characterization of naringinase enzyme and its application in debittering of Pomelo juice (*Citrus grandis*): A comparative study with macroporous resin." *Journal of Food Processing and Preservation* 45(5): 15380.

Gupta, R., Kumari, A., Syal, P., and Singh, Y. 2015. "Molecular and functional diversity of yeast and fungal lipases: Their role in biotechnology and cellular physiology." *Progress in Lipid Research* 57: 40–54.

Gupta, R., and Ramnani, P. 2006. "Microbial keratinases and their prospective applications: An overview." *Applied Microbiology and Biotechnology* 70: 21–33.

Gurung, N., Ray, S., Bose, S., and Rai, V. 2013. "A broader view: Microbial enzymes and their relevance in industries, medicine, and beyond." *BioMed Research International* 2013: 329121.

Hamer, R.J. 1995. "Enzymes in the baking industry." In *Enzymes in Food Processing*, edited by G.A. Tucker, and L.F.J. Woods. Springer, Boston, MA.

Hasan, F., Shah, A.A., Javed, S., and Hameed, A. 2010. "Enzymes used in detergents: Lipases." *African Journal of Biotechnology* 9(31): 4836–4844.

Hervey, G.W. 1925. "A Nutritional Study upon A Fungus Enzyme." *Science* 62: 247.

IDF 1990. "Int. Dairy Fed. Bull." 247: 24–38.

Ito, M., and Yamagata, T. 1989. "Purification and characterization of glycosphingo-lipid-specific endoglycosidases (endoglycoceramidases) from a mutant strain of *Rhodococcus* sp. Evidence for three molecular species of endoglycoceramidase with different specificities." *The Journal of Biological Chemistry* 264(16): 9510–9519.

Ito, S., Kobayashi, T., Ara, K., Ozaki, K., Kawai, S., and Hatada, Y. 1998. "Alkaline detergent enzymes from alkaliphiles: Enzymatic properties, genetics, and structures." *Extremophiles* 2(3): 185–190.

Itoh, T., and Hanefeld, U. 2017. "Enzyme catalysis in organic synthesis." *Green Chemistry* 19(2): 331–332.

James, J., Simpson, B.K., and Marshall, M.R. 1996. "Application of enzymes in food processing." *Critical Reviews in Food Science & Nutrition* 36(5): 437–463.

Janarthanan, R., Prabhakaran, P., and Ayyasamy, P.M. 2014. "Bioremediation of vegetable wastes through biomanuring and enzyme production." *International Journal of Current Microbiology and Applied Sciences* 3: 89–100.

Kamini, N.R., Hemachander, C., Mala, J.G.S., and Puvanakrishnan, R. 1999. "Microbial enzyme technology as an alternative to conventional chemicals in leather industry." *Current Science* 77: 80–86.

Kanth, S.V., Venba, R., Madhan, B., Chandrababu, N.K., and Sadulla, S. 2008. "Studies on the influence of bacterial collagenase in leather dyeing." *Dyes and Pigments* 76: 338–347.

Karam, J., and Nicell, J.A. 1997. "Potential applications of enzymes in waste treatment." *Journal of Chemical Technology & Biotechnology: International Research in Process, Environmental and Clean Technology* 69(2): 141–153.

Karnwal, A., Singh, S., Kumar, V., Sidhu, G.K., Dhanjal, D.S., Datta, S., Amin, D.S., Saini, M., and Singh, J. 2019. "Fungal enzymes for the textile industry." In *Recent Advancement in White Biotechnology through Fungi*, edited by Ajar Nath Yadav, Arti Gupta, Sangram Singh, and Shashank Mishra, 459–482. Springer, Cham.

Katoh, T., Katayama, T., Tomabechi, Y., Nishikawa, Y., Kumada, J., Matsuzaki, Y., and Yamamoto, K. 2016. "Generation of a mutant Mucor hiemalisendoglycosidase that acts on core-fucosylated N-glycans." *The Journal of Biological Chemistry* 291(44): 23305–23317.

Kavitha, K., Shankari, K., and Meenambiga, S.S.2021. "A review on extraction of lipase from Aspergillus Species and its applications." *Research Journal of Pharmacy and Technology* 14(8): 4471–4475.

Kenealy, W.R., and Jeffries, T.W. 2003. "Enzyme processes for pulp and paper: A review of recent developments." *American Chemical Society* 845: 210–239.

Khambhaty, Y. 2020. "Applications of enzymes in leather processing." *Environmental Chemistry Letters* 18(3): 747–769.

Kim, S., Lee, M.H., Lee, E.S., Nam, Y.D., and Seo, D.H. 2018. "Characterization of mannanase from *Bacillus* sp., a novel Codium fragile cell wall-degrading bacterium." *Food Science and Biotechnology* 27(1): 115–122.

Koeduka, T., Kajiwara, T., and Matsui, K. 2007. "Cloning of lipoxygenase genes from a cyanobacterium, Nostoc punctiforme, and its expression in *Eschelichia coli*." *Current Microbiology* 54(4): 315–319.

Kuhad, R.C., Gupta, R., and Singh, A. 2011. "Microbial cellulases and their industrial applications." *Enzyme Research*. doi: 10.4061/2011/280696

Kula, M.R. 1987. "Trends in enzyme technology." In: Hollenberg, C.P., and Sahm, H. (eds), *Biotec 1. Microbial Genetic Engineering and Enzyme Technology*, pp. 77. Gustav Fisher. New York.

Kumar, A., Bilal, M., Singh, A.K., Ratna, S., Rameshwari, K.T., Ahmed, I., and Iqbal, H.M. 2022. "Enzyme cocktail: A greener approach for biobleaching in paper and pulp industry." In: Bhat, R., Kumar, A., Nguyen, T.A., Sharma, S., (eds), *Nanotechnology in Paper and Wood Engineering*, 303–328. Elsevier.

Kumar, A., and Chandra, R. 2020. "Ligninolytic enzymes and its mechanisms for degradation of lignocellulosic waste in environment." *Heliyon* 6(2), e03170.

Kumar, A., and Singh, S. 2012. "Directed evolution: Tailoring biocatalysts for industrial applications." *Critical Reviews in Biotechnology* 33(4):365–378.

Kumar, C.G., Malik, R.K., and Tiwari, M.P. 1998. "Novel enzyme-based detergents: An Indian perspective." *Current Science* 75(12): 1312–1318.

Kumar, D., Kumar, A., Sondhi, S., Sharma, P., and Gupta, N. 2018. "An alkaline bacterial laccase for polymerization of natural precursors for hair dye synthesis." *3 Biotech* 8(3): 1–10.

Kumar, V., Khokhar, D., Sangwan, P., and Agrawal, S. 2012. "Role of Phytate and Phytase in human health." *Indian Farmers Digest* 45(5): 42–44.

Kumar, V., Sangwan, P., Singh, D., and Gill, P.K. 2014a. "Global scenario of industrial enzyme market." In *Industrial Enzymes: Trends, Scope and Relevance*, edited by Vikas Beniwal, and Anil Kumar Sharma 176–196. Nova Science Publishers, New York.

Kumar, V. Singh, D. Sangwan, P., and Gill, P.K.2014b. "Global market scenario of industrial enzymes." In: Beniwal, V., and Sharma, A.K. (eds), *Industrial Enzymes: Trends, Scope and Relevance*, pp. 173–196. Nova Science Publishers, New York.

Laachari, F., El Bergadi, F., and Ibnsouda, S.K. 2015. "Purification and characterization of a Novel Thermostable Lipase from *Aspergillus flavus*." *International Journal of Research* 2(2): 342–352.

Lalor, E., and Goode, D. 2010. "Brewing with enzymes." In: Whitehurst, R.J., and van Oort, M. (eds). *Enzymes in Food Technology*, 2nd edn. Chapter 8, pp. 163. Wiley.

Law, J., and Haandrikman, A. 1997. "Proteolytic enzymes of lactic acid bacteria." *International Dairy Journal* 7: 1–11.

Laxman, R.S., Sonawane, A.P., More, S.V.,et al. 2005. "Optimization and scale up of production of alkaline protease from *Conidiobolus coronatus*."*Process Biochemistry* 40: 3152–3158.

Li, G., Liu, X., and Yuan, L. 2017. "Improved laccase production by *Funalia trogii* in absorbent fermentation with nutrient carrier." *Journal of Bioscience and Bioengineering* 124(4): 381–385.

Lyons, T.P. 1991. *Biotechnology in the Feed Industry*. Alltech Technical Publications.

Madhavi, J., Srilakshmi, J., Rao, M.R., and Rao, K.R.S.S. 2011. "Efficient leather dehairing by bacterial thermostable protease." *International Journal of Bio-Science and Bio-Technology* 3: 11–26.

Madhu, A., and Chakraborty, J.N. 2017. "Developments in application of enzymes for textile processing." *Journal of cleaner production* 145: 114–133.

Malathi, S., and Chakraborty, R. 1991. "Production of alkaline protease by a new *Aspergillus flavus* isolate under solid-substrate fermentation conditions for use as a depilation agent." *Applied and Environmental Microbiology* 57: 712–716.

Mane, P., and Tale, V. 2015. "Overview of microbial therapeutic enzymes." *International Journal of Current Microbiology and Applied Sciences* 4(4): 17–26.

Martinez, A., and Maicas, S. 2021. "Cutinases: Characteristics and insights in industrial production." *Catalysts* 11(10): 1194.

Matama, T., Carneiro, F., Caparrós, C., Gübitz, G.M., and Cavaco Paulo, A. 2007. "Using a nitrilase for the surface modification of acrylic fibres." *Biotechnology Journal: Healthcare Nutrition Technology* 2(3): 353–360.

Mehta, A., Bodh, U., and Gupta, R. 2017. "Fungal lipases: A review." *Journal of Biotech Research* 8.

Mehta, A., Guleria, S., Sharma, R., and Gupta, R. 2021. "The lipases and their applications with emphasis on food industry." In *Microbial Biotechnology in Food and Health*, edited by Ramesh C. Ray, 143–164. Academic Press.

Mehta, P.K., and Sehgal, S. 2019. "Microbial enzymes in food processing." In: Husain, Q., Ullah, M. (eds), *Biocatalysis*, 255–275. Springer, Cham.

Meshram, A., Singhal, G., Bhagyawant, S.S., and Srivastava, N 2019. "Plant-derived enzymes: A treasure for food biotechnology." In *Enzymes in Food Biotechnology*, edited by Mohammed Kuddus, 483–502. Academic Press.

Mojsov, K. 2011. "Application of enzymes in the textile industry: A review." In: II International Congress "Engineering, Ecology and Materials in the Processing Industry", Proceedings, 2011, 09-11 March, pp. 230–239, Jahorina, Bosnia and Hercegovina.

Mokashe, N., Chaudhari, B., and Patil, U. 2017. "Detergent-compatible robust alkaline protease from newly isolated halotolerant *Salinicoccus* sp. UN-12." *Journal of Surfactants and Detergents* 20(6): 1377–1393.

Mukram, I., Nayak, A.S., Kirankumar, B., Monisha, T.R., Reddy, P.V., and Karegoudar, T.B. 2015. "Isolation and identification of a nitrile hydrolyzing bacterium and simultaneous utilization of aromatic and aliphatic nitriles." *International Biodeterioration & Biodegradation* 100: 165–171.

Nair, I.C., and Jayachandran, K. 2019. "Aspartic proteases in food industry." In: Parameswaran, B., Varjani, S., Raveendran, S. (eds), *Green Bio-processes*, 15–30. Springer, Singapore.

Nair, S.G., Sindhu, R., and Shashidhar, S. 2008. "Fungal xylanase production under solid state and submerged fermentation conditions." *African Journal of Microbiology Research* 2: 82–86.

Naveed, M., Nadeem, F., Mehmood, T., Bilal, M., Anwar, Z., and Amjad, F. 2021. "Protease – A versatile and ecofriendly biocatalyst with multi-industrial applications: An updated review." *Catalysis Letters* 151(2): 307–323.

Ni, H., Chen, F., Cai, H., Xiao, A., You, Q., and Lu, Y. 2012. "Characterization and preparation of Aspergillus niger naringinase for debittering citrus juice." *Journal of Food Science* 77(1): C1–C7.

Nisha, M., and Satyanarayana, T. 2016. "Characteristics, protein engineering and applications of microbial thermostable pullulanases and pullulan hydrolases." *Applied Microbiology and Biotechnology* 100(13): 5661–5679.

Niyonzima, F.N., and More, S.S. 2015. "Coproduction of detergent compatible bacterial enzymes and stain removal evaluation." *Journal of Basic Microbiology* 55(10): 1149–1158.

Ojha, B.K., Singh, P.K., and Shrivastava, N. 2019. "Enzymes in the animal feed industry." In *Enzymes in Food Biotechnology*, edited by Mohammed Kuddus, 93–109. Academic Press.

Osuji, A.C., Eze, S.O.O., Osayi, E.E., and Chilaka, F.C. 2014. "Biobleaching of industrial important dyes with peroxidase partially purified from garlic." *Scientific World Journal* 2014: 183163.

Ozatay, S. 2020. "Recent Applications of Enzymes in Food Industry."*Journal of Current Research on Engineering, Science and Technology* 6(1): 17–30.

Panda,T., and Gowrishankar,B.S. 2005. "Production and applications of esterases." *Applied Microbiology and Biotechnology* 67: 160–169.

Pandey, A., Nigam, P., Soccol, C.R., Soccol, V.T., Singh, D., and Mohan, R. 2000. "Advances in microbial amylases." *Biotechnology and Applied Biochemistry* 31(2): 135–152.

Pandey, D., Patel, S.K., Singh, R., Kumar, P., Thakur, V., and Chand, D. 2019. "Solvent-tolerant acyltransferase from *Bacillus* sp. APB-6: Purification and characterization." *Indian Journal of Microbiology* 59(4): 500–507.

Pandey, K., Singh, B., Pandey, A.K., Badruddin, I.J., Pandey, S., Mishra, V.K., and Jain, P.A. 2017. "Application of microbial enzymes in industrial waste water treatment." *International Journal of Current Microbiology and Applied Sciences* 6(8): 1243–1254.

Pandi, A., Ramalingam, S., Rao, J.R., Kamini, N.R., and Gowthaman, M.K. 2016. "Inexpensive α-amylase production and application for fiber splitting in leather processing." *RSC Advances* 6(39): 33170–33176.

Patel, A., Singhani, R.R., and Pandey, A. 2016. "Novel enzymatic processes applied to the food industry." *Current Opinion in Food Science* 7: 64–72.

Patel, A.K.Singhania, R.R., and Pandey, A. 2017. "Production, purification, and application of microbial enzymes." In *Biotechnology of Microbial Enzymes*, edited by Goutam Brahmachari, Arnold Demain, and Jose Adrio, 13–41. Academic Press.

Peng, Y.H., and Shih, Y.H. 2013. "Microbial degradation of some halogenated compounds: Biochemical and molecular features." In *Biodegradation of Hazardous and Special Products*, edited by R. Chamy, and F. Rosenkranz, 51–69.

Periolatto, M., Ferrero, F., Giansetti, M., Mossotti, R., and Innocenti, R. 2011. "Influence of protease on dyeing of wool with acid dyes." *Open Chemistry* 9: 157–164.

Peyrat, L.A., Tsafantakis, N., Georgousaki, K., Ouazzani, J., Genilloud, O., Trougakos, I.P., and Fokialakis, N. 2019. "Terrestrial microorganisms: Cell factories of bioactive molecules with skin protecting applications." *Molecules* 24(9): 1836.

Pham, L.T.M., Kim, S.J., and Kim, Y.H. 2016. "Improvement of catalytic performance of lignin peroxidase for the enhanced degradation of lignocellulose biomass based on the imbedded electron-relay in long-range electron transfer route." *Biotechnology for Biofuels* 9(1): 1–10.

Poulsen, P.B., and Klaus Buchholz, H. 2003. "History of enzymology with emphasis on food production." In *Handbook of Food Enzymology*, edited byJ.R. Whitaker, A.G.J. Voragen, and D.W.S. Wong. Marcel Dekker, New York.

Puri, M. 2012. "Updates on naringinase: Structural and biotechnological aspects." *Applied Microbiology and Biotechnology* 93: 49–60.

Rai, S.K., Roy, J.K., and Mukherjee, A.K. 2010. "Characterisation of a detergent-stable alkaline protease from a novel thermophilic strain *Paenibacillus tezpurensis* sp. nov. AS-S24-II." *Applied Microbiology and Biotechnology* 85(5): 1437–1450.

Rajkumar, R., Jayappriyan, K.R., and Rengasamy, R. 2011. "Purifcation and characterization of a protease produced by *Bacillus megaterium* RRM2: Application in detergent and dehairing industries." *Journal of Basic Microbiology* 51: 614–624.

Rathore, D.S.Sheikh, M., and Singh, S.P. 2021. "Marine Actinobacteria: New Horizons in Bioremediation." In *Recent Developments in Microbial Technologies*, edited by Chandrama Prakash Upadhyaya, Joginder Singh, Ram Prasad, and Vivek Kumar, 425–449. Springer, Singapore.

Raveendran, S., Parameswaran, B., Ummalyma, S.B., Abraham, A., Mathew, A.K., Madhavan, A., Rebello, S., and Pandey, A. 2018. "Applications of Microbial Enzymes in Food Industry." *Food Technology and Biotechnology* 56(1): 16–30.

Robinson, W.G., and Hook, R.H. 1964. "Ricinine nitrilase I. Reaction product and substrate specificity."*The Journal of Biological Chemistry* 239: 4257–4262.

Rozell, J.D., and Bommarius, A.S. 2002. *Enzyme Catalysis in Organic Synthesis*, 2nd ed., edited by K. Drauz, and H. Waldmann, vol. II, 873893. Wiley-VCH, Weinheim, Germany.

Sabu, A. 2003. "Sources, properties and applications of microbial therapeutic enzymes." *Indian Journal of Biotechnology* 2(3): 334–341.

Sadhasivam, S., Savitha, S., Swaminathan, K., and Lin, F.H. 2008. "Production, purification and characterization of mid-redox potential laccase from a newly isolated *Trichoderma harzianum* WL1." *Process Biochemistry* 43: 736–742.

Sanghi, A., Garg, N., Gupta, V.K., Mittal, A., and Kuhad, R.C. 2010. "One step purification and characterization of a cellulose free xylanase produced by alkalophilic *Bacillus subtilis* ASH." *Brazilian Journal of Microbiology* 41: 467–476.

Schafer, T.Kirk, O., andBorchert, T.V. 2002. "Enzymes for technical applications." In *Biopolymers*, edited by S.R. Fahnestock, and S.R. Steinbüchel, 377–437. Wiley-VCH, Weinheim, Germany.

Sharma, H.P., Patel, H., and Sharma, S. 2014. "Enzymatic extraction and clarification of juice from various fruits – A review." *Trends Post Harvest Technology* 2(1):1–14.

Sharma, P.K., and Chand, D. 2012. "Production of cellulase free thermostable xylanase from Pseudomonas sp. XPB-6." *International Research Journal of Biological Sciences* 1(5): 31–41.

Shi, K., Jing, J., Song, L., Su, T., and Wang, Z. 2020. "Enzymatic hydrolysis of polyester: Degradation of poly (ε-caprolactone) by *Candida antarctica* lipase and *Fusarium solani* cutinase." *International Journal of Biological Macromolecules* 144: 183–189.

Shiroya, A.J. 2021. "Recent advances of laccase enzyme" In *Industrial Biotechnology: A Review*. Pharma News.

Singh, D., and Chen, S. 2008. "The white-rot fungus Phanerochaete chrysosporium conditions for the production of lignin degrading enzymes." *Applied Microbiology and Biotechnology* 81: 399–417.

Singh, P., and Kumar, S. 2019. "Microbial enzyme in food biotechnology." In *Enzymes in Food Biotechnology*, edited by Mohammed Kuddus, 19–28. Academic Press.

Singh, P., Sulaiman, O., Hashim, R., Rupani, P.F., and Peng, L.C. 2010. "Biopulping of lignocellulosic material using different fungal species: A review." *Reviews in Environmental Science and Biotechnology* 9: 141–151.

Singh, R., Kumar, M., Mittal, A., and Mehta, P.K. 2016. "Microbial enzymes: Industrial progress in 21st century." *3 Biotech* 6(2): 1–15.

Singh, R., Singh, A., and Sachan, S. 2019. "Enzymes used in the food industry: Friends or foes?." In *Enzymes in Food Biotechnology*, edited by Mohammed Kuddus, 827–843. Academic Press.

Sirisha, V.L., Jain, A., and Jain, A. 2016. "Enzyme immobilization: An overview on methods, support material, and applications of immobilized enzymes." *Advances in Food and Nutrition Research* 79: 179–211.

Soccol, C.R., Rojan, P.J., Patel, A.K., Woiciechowski, A.L., Vandenberghe, L.P.S., and Pandey, A. 2005. "Glucoamylase." In *Enzyme Technology*, edited by A. Pandey, C. Webb, C.R. Soccol, and C. Larroche, 221–238. Asiatech Publishers, New Delhi, India.

Sogani, M., Mathur, N., Bhatnagar, P., and Sharma, P. 2012. "Biotransformation of amide using *Bacillus* sp.: Isolation strategy, strain characteristics and enzyme immobilization." *International journal of Environmental Science and Technology* 9(1): 119–127.

Solanki, P., Putatunda, C., Kumar, A., Bhatia, R., and Walia, A. 2021. "Microbial proteases: Ubiquitous enzymes with innumerable uses." *3 Biotech* 11(10): 1–25.

Souza, P.M.D., Bittencourt, M.L.D.A., Caprara, C.C., Freitas, M.D., Almeida, R.P.C.D., Silveira, D., Fonseca, Y.M., Ferreira, E.X., Pessoa, A., and Magalhães, P.O. 2015. "A biotechnology perspective of fungal proteases." *Brazilian Journal of Microbiology* 46: 337–346.

Srivastava, P.K., and Kapoor, M. 2014. "Cost-effective endo-mannanase from *Bacillus* sp. CFR1601 and its application in generation of oligosaccharides from guar gum and as detergent additive." *Preparative Biochemistry and Biotechnology* 44(4): 392–417.

Strohmeier, G.A., Pichler, H., May, O., and Gruber-Khadjawi, M. 2011. "Application of designed enzymes in organic synthesis." *Chemical Reviews* 111(7): 4141–4164.

Sukumaran, R.K., Singhania, R.R., and Pandey, A. 2005. "Microbial cellulases – Production, applications and challenges." *Journal of Scientific and Industrial Research* 64: 832–844.

Tauber, M.M., Cavaco-Paulo, A., Robra, H., and Gübitz, G.M. 2000. "Nitrile hydratase and amidase from Rhodococcus rhodochrous hydrolyze acrylic fibers and granular polyacrylonitriles." *Applied and Environmental Microbiology* 66: 1634–1638.

Thombre, R.S., and Kanekar, P.P. 2017. Cyclodextrin Glycosyl Transferase (CGTase): An overview of their production and biotechnological applications. In *Industrial Biotechnology: Sustainable Production and Bioresource Utilization*, edited by Rebecca S. Thombre, and Pradnya P. Kanekar, 141–159.

Tonin, F., Melis, R., Cordes, A., Sanchez-Amat, A., Pollegioni, L., and Rosini, E. 2016. "Comparison of different microbial laccases as tools for industrial uses." *New Biotechnology* 33(3): 387–398.

Torres, C.E., Negro, C., and Blanco, A. 2012. "Enzymatic approaches in paper industry for pulp refining and biofilm control." *Applied Microbiology and Biotechnology* 96(2): 327–344.

Trani, A., Loizzo, P., Cassone, A., and Faccia, M. 2017. "Enzymes applications for the dairy industry." *Industrial Enzyme Applications*, 166–175.

Uma, C., and Sivagurunathan, P. 2021. "Applications of fungal lipases – An overview." *Recent Trends in Modern Microbial Technology* 1: 195.

Urlacher, V.B., and Schmid, R.D. 2006. "Recent advances in oxygenase-catalyzed biotransformations." *Current Opinion in Chemical Biology* 10:156.

van Oort, M. 2010. "Enzymes in food technology – Introduction." In *Enzymes in Food Technology*, edited by Mohammed Kuddus, 2. Academic Press.

Vashist, P., Kanchana, R., Devasia, V.L.A., Shirodkar, P.V., and Muraleedharan, U.D. 2019. "Biotechnological implications of hydrolytic enzymes from marine microbes." In *Advances in Biological Science Research*, edited by Milind Naik, and Surya Nandan Meena, 103–118. Academic Press, London.

Vasic-Racki, D. 2006. "History of industrial biotransformations-dreams and realities", In *Industrial Biotransformations*, 2nd ed., edited by A. Liese, K. Seelbach, and C.S. Wandrey, 1–35. Wiley-VCH, Weinheim, Germany.

Verma, A., Singh, H., Anwar, S., Chattopadhyay, A., Tiwari, K.K., Kaur, S., and Dhilon, G.S. 2017. "Microbial keratinases: Industrial enzymes with waste management potential." *Critical Reviews in Biotechnology* 37(4): 476–491.

Verma, A.K., Soniya, S., Nishad, S., Vinod, K., Shalini, S., and Ashutosh, D. 2013. "Production, purification and characterization of β-glucosidase from *Bacillus subtilis* strain PS isolated from sugarcane bagasse." *Journal of Pure and Applied Microbiology* 7(1):803–810.

Vulfson, E.N.1994. *Industrial Applications of Lipases in Lipases*, edited by P. Wooley, and S.B.D. Petersen, 271. Cambridge University Press, Cambridge, Great Britain.

Walia, A., Guleria, S., Mehta, P., Chauhan, A., and Parkash, J. 2017. "Microbial xylanases and their industrial application in pulp and paper biobleaching: A review."*3 Biotech*, 7(1): 1–12.

Walsh, G.A., Power, R.F., and Headon, D.R. 1993. "Enzymes in the animal-feed industry." *Trends in Biotechnology* 11(10): 424–430.

Wavhal, S.D., and Balasubramanya, R.H. 2011. "Role of biotechnology in the treatment of polyester fabric."*Indian Journal of Microbiology* 51(2): 117–123.

Weyler, W., Hsu, Y.P.P., and Breakafield, X.O.1990. "Biochemistry and genetics of monoamine oxidase." *Pharmacology & Therapeutics* 47: 391–417.

Wilson, D.B. 2009. "Cellulases and biofuels." *Current Opinion in Biotechnology* 20(3): 295–299.

Winkler, M., Geier, M., Hanlon, S.P., Nidetzky, B., and Glieder, A. 2018. "Human enzymes for organic synthesis." *Angewandte Chemie International Edition* 57(41): 13406–13423.

Xia, W., Zhang, K., Su, L., and Wu, J. 2021. "Microbial starch debranching enzymes: Developments and applications." *Biotechnology Advances* 50: 107786.

Yin, S.J., Han, C.L.Lee, A.I., Wu, C.W. 1999. "Human alcohol dehydrogenase family. Functional classification, ethanol/retinol metabolism, and medical implications." *Advances in Experimental Medicine and Biology* 463: 265–274.

Yokoyama, K., Nakamura, N., Seguro, K., and Kubota, K. 2000. "Overproduction of microbial transglutaminase in *Escherichia coli*, in vitro refolding, and characterization of the refolded form." *Bioscience, Biotechnology, and Biochemistry* 64: 1263–1270.

Young, L., andYu, J. 1997. "Ligninase-catalysed decolorization of synthetic dyes." *Water Research* 31: 1187–1193.

Zafar, A., Aftab, M.N., Asif, A., Karadag, A., Peng, L., Celebioglu, H.U., Afzal, M.S., Hamid, A., and Iqbal, I.2021. "Efficient biomass saccharification using a novel cellobiohydrolase from *Clostridium clariflavum* for utilization in biofuel industry." *RSC Advances* 11(16): 9246–9261.

Zaidi, K.U., Ali, A.S., Ali, S.A., and Naaz, I. 2014. "Microbial tyrosinases: Promising enzymes for pharmaceutical, food bioprocessing, and environmental industry." *Biochemistry Research International* 2014: 854687.

Zhu, D.,Wu, Q.N., and Wang, N. 2011. "Industrial enzymes." In *Comprehensive Biotechnology*, 2nd ed., vol. 3, 3–13. Chinese Academy of Sciences, Tianjin, China.

3

Optimization of Fermentation Process: Influence on Industrial Production of Enzymes

Ajay Nair and Archana S. Rao
Dayananda Sagar University, Bangalore, India

S. M. Veena
Sapthagiri College of Engineering, Bangalore, India

Uday Muddapur
KLE Tach University, Hubli, India

K. S. Anantharaju and Sunil S. More
Dayananda Sagar University, Bangalore, India

CONTENTS

DOI: 10.1201/9781003292333-3

3.1 Introduction

3.1.1 On the Production of Enzymes

Biocatalytic enzymes are foundational to the modern biotechnology industry to serve man's pursuit of utilizing living systems for several commercial purposes. Some of these purposes such as winemaking and brewing could be traced back as far as the dawn of history. Enzymes equal fermenting yeast cells that mediated the transformation of substrates to desirable products such as wine. The initial conclusive signs showing the potential of enzymes in modern industry came from an intriguing observation where the alcohol precipitates from extracts of malt contained a heat-sensitive biomolecule that produced fermentable sugars from starch. This enzyme was later termed diastase for its ability to convert starch into soluble dextrins. Several other enzymes such as peroxidases, pepsins, invertases, etc., were identified toward the later part of the nineteenth century. An amylolytic enzyme called Taka-Diastase produced by *Aspergillus oryzae* was the first enzyme to be patented (Headon and Walsh 1994).

From the moment industrialists realized the importance of enzymes, especially in modulating biotechnological industrial processes, living systems have constantly been explored for identifying several potential enzymes. In recent times, several applications ranging from the production of industrial materials to diagnostic commodities, therapeutic agents, etc., are heavily reliant on enzymes (Aftab et al. 2018). Although there are multiple biological sources, most enzymes are primarily derived from microorganisms via fermentation processes (Wiseman and Gould 1971; Fogarty 1983; Rose 1980).

The choice of source is key to designing the production workflow for enzymes. Plants, animals, and microorganisms are the possible sources of enzyme production. One of the fundamental reasons behind using microorganisms for the synthesis of commercial enzymes is the inherent ability of these simple organisms to produce large quantities of the biomolecule in a short period of time. The fermentation processes in recent times have been well optimized to synthesize enzymes in abundance and on demand (Fogarty 1983). Additionally, microbially derived enzymes are relatively more stable than those obtained from animal and plant sources, even when stored in optimal conditions for an extended period of time.

Another advantage of enzymes classically produced from microbial sources is being largely extracellular in nature. Extracellular enzymes make the further downstream processing steps simpler necessitating only a separation of the producer cells from the extracellular medium by either centrifugation or filtration. Later, the enzymes are merely precipitated out from the media. Proteases and amylolytic enzymes are the usual extracellular enzymes produced via most fermentation processes, in addition to pectinolytic and cellulolytic enzymes (Schwardt 1990).

In comparison to microbes, plants may not seem to be convenient sources of industrially potential enzymes. However, there are several enzymes of interest that have been obtained from plant sources. The limitations associated with deriving potential proteins of interest from plants range from seasonal factors, the nature of plant growth, or the fact that most plant-based enzymes are intracellular. But despite such shortcomings, some key enzymes such as proteases are among the most sought after industrial enzymes (Moffat 1992). Many significant enzymes of great applicability have been obtained from animal sources, particularly those from urine or snake venom. Rennin or chymosin is one of the industrially significant enzymes finding use in cheese production, and in the production of digestive aid. Renin was traditionally obtained from the stomach of suckling calves by extracting strips and incubating the latter in a 5–10% saline solution containing boric acid. Rennet is an extract, obtained from rennin, comprising several proteolytic and other enzymes. These enzymes within this extract have not been completely characterized but they do seem to hold immense catalytic potential. However, the availability of rennet purely relies on sacrificing young calves. Some of the other enzymes derived from animal sources are of therapeutic potential. For example, Ancrod is a serine protease derived from the venom of the Malaysian pit viper (Headon and Walsh 1994). This enzyme mediates the degradation of blood fibrin before they form clots. So, this demonstrated the ability of Ancrod to act as a therapeutic anticoagulant. Another serine protease is Urokinase which has been used widely as a thrombolytic agent for degrading clots (Ratzkin 1981). Urokinase is produced by the human kidney and is purified from human urine by subjecting the biological sample to a series of gel and ion-exchange chromatographic analyses.

By coupling mutational studies to thorough screening and selection techniques, industries have managed to develop microorganisms capable of secreting animal-specific enzymes in copious amounts. However, abundant quantities of enzymes produced via recombinant DNA technology are not significantly larger than what is usually produced by cells naturally. This particularly holds true with enzymes of therapeutic value. But there is an alternate reason for using genetic engineering for the production of such low-yielding proteins. The possibility of the accidental spread of disease through infected animal samples is reasonably high. And using recombinant DNA technology to enable organisms with the ability to synthesize the same protein avoids the aforesaid issue. For example, tissue plasminogen activator (tPA) has been shown to possess thrombolytic properties. Gene manipulation techniques have been applied to express this protein in multiple hosts. One of the most popular hosts was expressing it in the milk of transgenic mice. A recombinant DNA containing a promoter, regulatory sequence from the mouse whey acidic protein gene fused to the human tPA gene sequence (Gordon 1987). Genetic engineering has allowed for the high production capabilities of enzymes across a wide range of animal hosts. Genetically engineered human DNAse has shown to be effective in the treatment of cystic fibrosis (Collins 1992). Cystic fibrosis is typified by the production of extreme viscous mucous that disrupts lung function. The mucous is generated due to an immune response against bacterial infection. The large quantities of extracellular microbial DNA released mixed with other lung constituents form the highly viscous mucous in patients. Traditionally, this mucous was dislodged and aspirated out by physically thumping on the patient's chest. Inhaling the recombinant DNAse via nebulization promotes enzyme-driven degradation of the pathogenic DNA. Thereby significantly reducing the viscosity of the mucous in patients (Edgington 1993). Recombinant DNA technology has been applied to other industrially relevant enzymes such as amylases and proteases. For example, alpha-amylases gene from *B. amyloliquefaciens* has been expressed in *S. cervisiae*. The said enzyme is secreted into the extracellular media and later purified, for its ability to degrade starch (Svensson and Sogaard 1992). Alpha-amylases are among the most commonly used enzymes in starch processing. The recombinant enzyme was shown to be superior in terms of thermal stability and the ability to stay operational over a wide pH range (Pen 1992).

In the majority of cases, although microbes represent the prime source for the production of enzymes, only a finite number of microorganisms are considered to be industrially important. The reason behind only a limited number of microbes contributing to the global demand for biotechnologically relevant enzymes is that they must abide by certain critical norms such as being generally recognized as safe (GRAS); they should be non-pathogenic and non-toxic; also, they shouldn't typically produce antibiotics. Species of *Bacilli* and *Aspergilli* are the most commonly used GRAS-listed enzyme producers in the industry (Lambert and Meers 1983).

The systematic process of producing microbial enzymes begins by screening for the ideal candidate microorganisms as per the GRAS norms. Preliminary screening involves culturing candidate organisms on agar plates saturated with the substrate for the enzyme of interest (Howson and Davis 1983). Confirmation of enzyme secretion is made based on the detection of substrate hydrolysis zones. The initial candidate microbe screening is, however, not sufficient as it yields several potential producers. Additional technical assays downstream help in the selection of high-yielding producer microbes. Further optimization of the selection process includes scrutinizing and regulating the fermentation conditions (McNeil and Harvey 1980). Generally, the entire process leading up to the isolation of producer microorganisms with the desired quantity and type of the specific enzyme takes several years. Advancements in recombinant DNA technology have offered some despite developing hyperproducing microbes.

3.1.2 Fermentation Technology in Enzyme Production

Fermentation technology as an approach uses microbes for the production of not just enzymes but many other bioactive compounds that find purpose in industries such as energy, material, pharmaceutical, food, etc. Despite the long-standing success, this technology has had in meeting the demands for sustainable production of compounds, efforts have been put in to make strides in the design of processes and technologies for the process of fermentation (Campbell-Platt 1994). Primarily two methods are used for the commercial production of enzymes: submerged and solid-state fermentation (SSF) (Table 3.1).

3.1.2.1 Submerged Fermentation

This type of fermentation involves culturing microbes in a nutrient broth medium. Temperature, pH, and agitation are maintained at standardized levels for higher production. Selected microorganisms (bacteria, fungi) are grown in a broth of nutrients in enclosed vessels. Potential enzymes are produced in response to microbes utilizing the nutrients in the media. Typically, the vessels used can contain volumes of up to 1000 cubic meters. Most industrial enzymes produced by the microbes are expelled into the fermentation media to break down carbon and nitrogen raw materials such as maize, sugar, soya, etc., as far as the fermentation processes are concerned in any fermentation technology, there are two most common. A batch-fed process wherein liquid nutrients are introduced to the fermenter in accordance with the growth of the biomass. On the contrary, a continuous process involves continuously feeding the fermenter with media at the same rate as the used fermentation broth leaves the system. This ensures that growing cells receive a constant supply of fresh medium and that the products and effluents are simultaneously removed. This type of setup also permits the fermenter to be used for extended periods of time. Despite the multitude of factors that make continuous fermentation superior to batch fermentation, these reactors have not been used across a wide range of industries. And some of the prime reasons accounting for this are proclivity to get easily contaminated and performance glitches, such as those involving non-linear processes that are difficult to regulate (Renge et al. 2012).

TABLE 3.1

Industrially Relevant Enzymes and Their Sources of Origin

Enzyme	Source	Fermentation Process Employed	Industrial Use
Proteases	*B. subtilis,* *A. oryzae,* *Streptomyces* spp.	Submerged fermentation	Hydrolytic proteins widely used in baking, leather, and detergent industry
Amylases	*Bacilli* spp.	Solid-state fermentation	Starch liquefaction
Invertases	*Saccharomyces* spp.	Solid-state fermentation	Hydrolytic enzyme breaks sucrose to yield simple sugars
Penicillin acylase	*E. coli, Streptomyces* spp.	Recombinant DNA technology combined with Submerged Fermentation	Synthesis of semi-synthetic penicillin
Pectinases	*A. niger*	Solid-state fermentation, Submerged Fermentation	Hydrolytic enzymes breaking pectin, widely used in fruit juice production
Chymosin	Calf stomach	Recombinant DNA technology combined with submerged fermentation	Use as a thrombolytic agent
Glycosyltransferases	*Bacilli* spp.	Batch and fed-batch fermentation	Used for the production of cyclodextrin from starch

3.1.2.2 Solid-State Fermentation

SSF uses an alternate approach for the production of industrial enzymes. Usually, the extraction of enzymes from fungal sources follows surface fermentation techniques. This is also called Koji fermentation or tray cultivation or thin layer cultivation. Commonly paper pulp and agricultural wastes like wheat bran, rice bran, bagasse, etc., are used as substrates. There are multiple factors like pH, temperature, aeration, and humidity that affect the successful production of enzyme by fermentation (Renge et al. 2012). As opposed to submerged fermentation, solid state fairs better in some areas, such as higher volume for production, greater product yield with minimal effluent generation, and most importantly necessitates simpler equipment. The choice of substrate for SSF chiefly depends on the cost and its availability. Substrate selection also depends on the size and moisture levels of the compounds. Small substrates although offer larger surface areas for microbial growth, too small a size could potentially hamper cellular respiration. This in turn could impede microbial growth and essentially enzyme production yield. On the other hand, larger particles compromise the cell surface area but tend to provide better aeration for growth and enzyme production. Because SSF needs the substrate to be saturated with water, the moisture levels need to be standardized to support optimal microbial growth. Some of the industrially relevant enzymes produced by SSF include proteases, pectinases, glucoamylases, cellulases, etc. (Suganthi 2011).

3.2 Operational Issues with Fermentation Process Engineering

When compared to simple chemical processes, process engineering involved in the production of bioprocesses such as enzymes require accounting for all elements related to process modeling, design, and choice of control strategies. Therefore, much attention has been diverted toward standardizing control techniques in industries (Henson 2003). Typically, fermentation bioprocess modeling is categorized into two approaches: first, unstructured kinetics, and second structured kinetics. Biochemical processes generally rely on unstructured kinetic modeling. This is a non-mechanistic method that assumes that inhibition of product formation takes a linear, exponential curve. In other words, the kinetics of enzyme secretion relies on substrate concentration, substrate type, and the type of microbe. Although this model fits well in most experimental data, it has demonstrated inferior potential in modeling complex bioprocesses such as microbial enzyme production at industrial scales. This has been attributed to a few computational issues that go unaddressed (Moser and Starzak 1994). On the other hand, structured kinetics modeling recognizes cellular processes into distinct reaction patterns:

(a) catabolic pathways wherein substrates are broken down into energy and smaller molecules
(b) anabolic pathways wherein biomass precursors are synthesized
(c) multiplication of precursors to form biomass
(d) overall maintenance measures to keep the bioprocess operational

(Starzak et al. 1994)

The structured model in comparison to the unstructured model fairs better as it is able to account for all the intermediate biochemical reactions that lead up to enzyme production or product in general. Also, the inability of the unstructured model to appreciate how environmental factors affect a certain branch point in the metabolic process. A branch point in metabolism helps describe the extent to which the provided substrate is directed toward

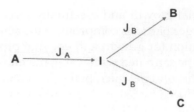

FIGURE 3.1
Metabolic branch point.

the expected product. Branch point kinetic modeling also predicts the rate at which substrate converts to a product (Figure 3.1).

In Figure 3.1, A represents the substrate, B represents the product, and C represents the byproduct; I represents the metabolite, and J represents the flux. To improve the formation, one would need to minimize the flux of C (J_C). Meanwhile, the flux J_B needs to be increased by directing more carbon toward B. This alteration of flux around metabolite I would need a thorough understanding of the entire metabolic network (Bideaux et al. 2006), or, in other words, the structured model kinetics. Structured model kinetics was applied to the production of ethanol from glucose where glycerol is the alternate byproduct. Interestingly, the model helped in significantly reducing the formation of glycerol during ethanol production.

3.2.1 Fermentation Control Specifics

Some of the major challenges in modulating fermentation processes include the following:

a. Availability of suitable sensors at critical checkpoints of the entire process
b. Selection algorithms suited for multiple fermentation controls.

For example, the model is incapable of predicting variables such as microbial cell concentration and viability. There are several other input-output variables that simply cannot be accounted for by an unstructured kinetic model, making fermentation processes so challenging. And this is despite the advances made in biosensor and digital communication technology. This has therefore encouraged research to explore better ways of controlling and interpreting fermentation processes. Among the most widely investigated options include adaptive control techniques, neural networks, fuzzy and model-based controls, etc. (Arbel et al. 1996; Schügerl 2001). Neural networks in comparison to many other structured or unstructured kinetic models do not rely on knowing how different process variables are connected to each other and the process in general. To understand how effective neural networks are in predicting biomass and substrate concentrations, the performance was compared with those derived using the typical Extended Kalman Filter (EKF). EKF on its own is quite efficient in optimizing fermentation process optimization. This algorithm successfully predicts aspects of cellular growth and concentration online. But when compared with neural networks, they were unable to comprehensively predict the structure of the dynamic fermentation process with great accuracy. Despite the overarching success neural networks have had, they are still in their initial stage.

3.3 Factors Affecting Fermentation Process

Of all the causes that could potentially determine the successful production of industrial enzymes, factors affecting the process of fermentation have the most impact. Chief factors modulating the production of enzymes via fermentation include both abiotic and biotic factors like media composition, temperature, pH, mechanical stress, degree of aeration, and the overall microbial cell morphology. Each and every factor is optimized individually and then the synergistic effect is studied (Flowchart 3.1).

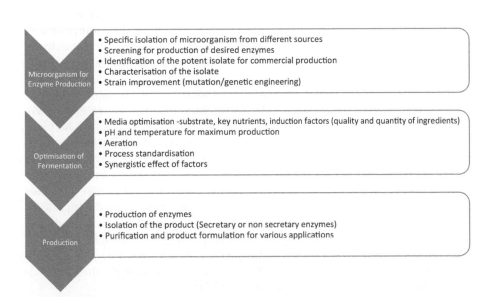

FLOWCHART 3.1
Steps involved in microbial production of enzymes.

3.3.1 Media Composition

Media standardization plays a seminal role in enzyme production by fermentation. The chemical composition of media has a direct impact on overall production. Any media should provide basic requirements like energy source, carbon, nitrogen, growth factors, and stimulating factors for microbial growth and enzyme production. Optimization of media composition not only helps in maximizing production but also helps in preventing the wastage of components which are unspent. The media cannot be generalized to bacteria and fungi. The components of the media, the induction factors, and key requirements are strain-specific. A high microbial growth rate will not always imply high enzyme production. The growth can be supported by carbon, nitrogen, and energy sources of the media; at the same time, there are factors that induce or inhibit the enzyme production (Ire 2011; Kumar and Takagi 1999).

The growth media optimization is normally carried out by one parameter at a time method or statistical methods. Largely, bacteria and fungi are used for the industrial production of enzymes. The specificity of production and stability of the product are very important factors in selecting the microorganism for fermentation. Enzymes produced by microorganisms can be either intercellular or secretary in nature. For industrial production, secretary enzymes are always a great choice, as it makes the isolation and purification easy. Genetic modification or fusion protein technique can be efficiently used to make the enzyme secretary. Genetically modified microorganisms are also used frequently to enhance enzyme production or sometimes, to suit the fermentation conditions. The fermentation conditions like temperature, pH, and aeration have to be less energy-requiring to set up fermentation. Induction of enzyme production by microbes requires activating factors which are included in the media.

The substrate used for fermentation, especially SSF, is agro-industrial wastes or byproducts. This makes the fermentation process less expensive, around 30% can be reduced. The utilization of such substrates also helps in bioremediation. Wheat bran, chicken feathers, Coco pith, tea stalks, sawdust, rice bran, organic kitchen wastes, etc., are some such inexpensive substrates used for the production of enzymes like amylase, invertase, keratinase, laccase, tannase, xylanase, cellulose, and many more (Table 3.2). On the other hand, submerged fermentation requires expensive synthetic media for microbial growth and production. Some enzymes like invertase can be produced by both types of fermentation by *Aspergillus* spp. In some cases, for example, enzymes produced for detergents, are produced synchronously by a single production medium. In such cases, the medium should contain the stimulating factors for all the enzymes (Niyonzima 2019).

The concentration of nutrients in the media will have a huge impact on the production of enzymes. Growth and metabolic activity depend on both the quality and quantity of the media ingredients. And hence, media

TABLE 3.2

Commonly Used Media Components for Microbial Enzyme Production

Carbon sources/Energy sources	Molasses, malted barley, starch, bagasse, paper pulp, oils, fats, corn steep liquor, wheat bran, rice bran, etc.
Nitrogen sources	Corn steep liquor, yeast extract, peptone, soya bean meal, etc.
Minerals	Cobalt, copper, iron, manganese, molybdenum, zinc, etc.
Chelators	EDTA, citric acid, polyphosphates, etc.
Vitamins/growth factors	Fe, Cu, Mn, Mo, Vit B, etc.
Precursors	Tryptophan, leucine, geraniol, citral
Inducers	Enzyme specific (e.g., Cellulose, cellobiose, cellobiose diplamitate for cellulase; dextrans, isomaltose, isomaltose diplamitate for dextranase; sucrose, sucrose monopalmitate for invertase)
Inhibitors	Formic acid, acetic acid
Water	Distilled/mineralized water
Antifoaming agent	Castor oil, silicon derivatives, glycol, etc.

optimization should aim for minimal microbial growth with maximum product formation. This can minimize the drawbacks like byproduct generation, toxin accumulation, and high cost (Singh et al. 2017).

3.3.2 pH

pH is a main factor affecting the growth of microorganisms. The microbial growth, fermentation rate, and byproduct formation are affected by pH changes. Even the slightest fluctuation from the optimum could potentially lead to poor microbial growth and low product yield. For example, β-glucosidase production from *Trichoderma viride* optimally occurs at a pH of 5.5. Any deviation from the norm, either more acidic or more basic did not permit microbial growth or production of the enzyme (Ikram-ul-Haq et al. 2006). Not only does an non-optimal pH affect growth, but it has been demonstrated in several species of *A. niger* that unfavorable pH could hamper the production and the secretion of enzymes. Optimal pH does this by limiting nutrient accessibility (Bajpai 1997). Production of some of the most widely used enzymes such as amylases may be both pH-dependent and pH-independent, depending on the microbe. Amylase produced by *A. oryzae* required a pH of 7–7.5. On the other hand, *A. ochraceus* produced amylase around a pH of 3–6 (Nahas and Waldemarin 2002).

3.3.3 Temperature

No other ambient condition affects microbial growth and the subsequent enzyme yield as much as temperature. Therefore, to achieve high product yield in proportion to biomass growth, providing the apt temperature is critical. Temperature cultivation can be offered via three different approaches:

(a) carrying out fermentation at a single and constant temperature
(b) conducting fermentation in two stages at separate temperatures
(c) fermentation in dynamic temperature shifts.

(Zheng et al. 2001; Zhang et al. 2002)

The effect of temperature has been demonstrated specifically on the microbial response to enzyme concentration; microbial growth rate, fermentation rate, cell viability, duration of the lag/log phase, etc. (Jackson 2000). An alternate study looking at enzyme production by yeasts, found that increasing temperature from the optimal led reduction in lag phase, increase in the net growth, and conversion of carbon to biomass. The overall product yield decreased thus, dramatically (Sener and Canbas 2007). Fermentative production of the enzyme elastase has been shown to be maximal when cultures of Bacilli spp. were grown at a constant temperature of 30°C. it has been also observed that the overall fluctuation in the operational parameters of a reactor such as temperature, rate of media flow, substrate concentration, etc., could potentially influence enzyme production. For example, the production of xylanase by *A. niger* seems to vary in its ability to produce xylanase. When the reactor reached higher temperatures, there was increased biomass with compromised enzyme production. However, when the temperature dropped to 27°C, the net production time reduced from 92 to 76 hours, without impacting the production of xylanase (Yuan et al. 2005).

3.3.4 Mechanical Forces and Aeration

In a bioreactor, the strength of agitation and oxygenation hugely impacts the morphology of the microbe. The physical stress caused due to mechanical aeration has been shown to damage mycelia. This is why it is necessary to properly standardize the agitation rate to uphold cell viability. Recent work on the production of lipase via submerged fermentation has consistently demonstrated that improved oxygen transfer rate through agitation and perfluorocarbon carriers maximized enzyme production (Frost and Moss 1987). Perfluorocarbons have the potential to increase the solubility of oxygen, thereby improving the net oxygen transfer to cells (Riess and LeBlanc 1982). An alternate study, investigating the impact of oxygenation on the production of lipase by *Rhizopus arrhizus* comparisons, was drawn between two modes of enzyme production: one with regulated oxygen concentration and the other with regulated aeration rate. It was observed that the rate of oxygenation influenced the production of the enzyme more than microbial growth (Murat and Dursun 2000). The effect of aeration on *A. oryzae* was studied, and it was noticed that the reduced enzyme productivity was due to changes in cell morphology, lowered biomass, and insufficient mixing due to inappropriate aeration (Amanullah et al. 1999). Agitation rates beyond the optimal levels cause the cells to form small, dense pellets. The formation of compact pellets is induced by first the pellicles being shaved off from the pelleted surface, and second due to the disruption of the pellet at stronger agitation (Taguchi et al. 1968).

3.4 Optimization of Fermentation Process Technology

In any fermentation process, the medium of culture and the overall process at a physiological and biochemical level play a seminal role. This is because both these aspects determine the overall concentration and yield of the desired product. Therefore, it is considered essential to rigorously take optimization measures toward finding optimal medium and optimal process (Schmidt 2005). However, process optimization is not as straightforward as it might seem, and it has its fair share of challenges, be it being labor-intensive, expensive, time-consuming requirements for setting up many experiments. Setting up for multiple standardization experiments is vital simply due to the overabundance of mutant strains being constantly introduced into the microbial population. Every optimization process thus involves standardizing different combinations of process conditions and media components with their respective product output (Stanbury et al. 1997).

Some of the widely adopted optimization strategies applied to the fermentation process include a close-ended system and an open-ended system. In the close-ended system, the number and type of parameters assessed are known or fixed. But the limitation is that there are a known number of permutations and combinations that can be explored. On the contrary, an open-ended system can explore umpteen different parameters and combinations for optimizing the fermentation process. Additionally, an open-ended system is considered more suitable because it is unbiased making no presumption on which parameter component is potentially ideal for a said fermentation process (Kennedy and Krouse 1999).

3.4.1 Literature Mining

In this method, the conditions and components of the medium are accessed via literature survey, including the genus and species or microbial strain used. But the problem with this method is the overabundance of options. Consequentially one needs to shortlist the number and type of parameter/components. However, compared to some of the other methods, this is a fairly simple method as it does not require any prior mathematics skill (Kennedy and Krouse 1999).

3.4.2 Nutrient Swapping

This open-ended optimization strategy only compares components of one type in a medium. The method merely tests out the effects of replacing one component from the medium with an alternate one. Since the optimization strategy does not assess the variability due to component interactions, it could be considered a basic strategy. Nonetheless, the method does offer some positives, such as aiding with the screening of different carbon and nitrogen sources toward the overall standardization of the fermenting medium. For example, optimal carbon and nitrogen sources were screened for the production of mevastatin and citric acid by using the component replacement technique (Ahamad et al. 2006).

3.4.3 Biological Simulation

This is a close-ended system for fermentation process optimization. Despite being a closed system, this method allows for the optimization of several components of the media in addition to assessing cell growth (on the basis of mass balance strategy). But one of the major limitations of using this method is it being uneconomical, laborious, and consuming a significant amount of time. But one advantage of this method is that it does give an insight into the effects of different micro- and macro-components of the media on microbial growth (Kennedy and Krouse 1999).

3.4.4 One Factor-at-a-Time

This method adopts the close-ended strategy for optimizing fermentation processes. It is applicable to both the standardization of media components and parameters of the fermentation process. This method is premised upon the strategy of experimenting with one variable while keeping other factors constant (Alexeeva et al. 2002). Although it's a fairly simple strategy, it is time-consuming and expensive when there are too many variables to be tested. Also, the method ignores interactions between different components. Regardless of some of these shortcomings, one-factor-at-a-time is the most popular close-ended strategy employed toward fermentation process optimization (Kumar et al. 2003).

3.4.5 Factorial Design

In a close-ended system, in comparison to the one-factor-at-a-time method, each variable, be it from the medium or process conditions, can be independently modulated at more than one level. In other words, the assessment is performed in a factorial design with utmost efficiency. This method optimizes by first designing a factorial equation that describes a particular fermentation yield as a function of the fermentation process. The equation helps with making predictions on the effects of individual factors and the consequent interaction. A full factorial design tests out every factor and every possible combination of factor levels. Some of the most commonly assessed factors include culture strain type, media components, temperature, humidity, reaction pH, the total volume of the inoculum, etc. (Tunga et al. 1999).

3.4.6 Plackett and Burman's Strategy

This strategy serves its purpose when there are more than five variables to be investigated (Naveena et al. 2005). The technique there assesses n variables by n+1 experiment, as long as the n is a multiple of 4. Any factor that doesn't get accounted for as a variable is termed a dummy variable. Dummy variables are used as controls to eliminate the effect of variance (Plackett and Burman 1946).

3.4.7 Response Surface Methodology

Response Surface Methodology (RSM) is a statistical design that predicts fermentation process outcomes from a minimum number of standardizing experiments. RSM helps in identifying not only the significant factors but also the level the factor needs to be at for maximum product yield (Sayyad et al. 2007). Mathematical models developed by RSM successfully describe the interaction between dependent and independent variables for a fermentation process under investigation.

3.4.8 Evolutionary Operation

This optimization approach relies on a factorial-based design to help standardize the product yield. Alterations to any variable can be only made to a given cycle and not across cycles. This is to ensure that the predicted increase in product generation is greater than the estimated experimental error. This particular method was used to optimize the production of protease enzymes from *R. oryzae* (Banerjee and Bhattachaaryya 1993).

An upgraded optimization method, called Evolutionary Operation (EVOP) factorial design is a hybrid of the evolutionary operation and the factorial design method. So, the experimental variables are designed using the factorial technique and the results are analyzed using the EVOP. This strategy primarily involves multivariable sequential searching, wherein the effects of each variable can be compared with that of the other, and the overall interaction can be assessed. For each variable analyzed, the response is measured statistically. For example, in a study of four variables, the total number of new experiments that can be performed and investigated is 2^4. This number is excluding the control experiments. An optimized protocol for the production of gallic acid and protease from *R. oryzae* under SSF was designed using the EVOP-factorial technique (Kar et al. 2002).

3.4.9 Artificial Neural Network

When there is a large amount of data variables to work with, artificial neural networks (ANN) work best without needing any mechanistic description of the experimental system. And this becomes particularly useful toward optimizing media components (Kennedy and Krouse 1999). ANN operates by initially constructing a series of trial experiments. Following this, a network is created to connect data points. As the data points (either media composition or process condition) are connected, the system learns about the interaction between different data points. Based on its learning, ANN makes a prediction on the output in the form of both microbial performance and product (enzyme) production. Because of its ability to connect experimental variables and predict their interaction, ANN standardizes the fermentation process by taking the least amount of time (Patnaik 2005).

3.4.10 Fuzzy Logic

This optimization technique is premised upon a set of rules to optimize the fermentation process. The rules fundamentally range from defining the level (high or low) of components in a fermentation media to the minimum number of components, to understanding synergy between components (Ul-haq and Mukhtar 2006). Once a medium is entered into the fuzzy logic program, it is capable of predicting the performance of a particular strain and its product outcome (Anderson and Jayaraman 2005).

3.4.11 Genetic Algorithm

Amidst other optimization tools, genetic algorithm (GA) is a non-statistical based technique. Without requiring empirical models, this stochastic search tool optimizes the fermentation process remarkably well. While standardizing fermentation media components or fermentation process conditions, GA assumes each fermentation media constituent to be like a gene encoded in a chromosome. After completing the first series of experiments, the chromosome with maximum productivity is selected. Multiple chromosomes of the said kind are compared to get the best strain and the best fermentation condition for the desired product. the only shortcoming of this technique is its inability to store information after every optimization stage (Zuzek et al. 1996). Consequentially it is believed that using GA in combination with other strategies such as RSM, EVOP, and ANN. This ensured that the fermentation processes are standardized optimally toward product output with maximum yield (Figure 3.2) (Nagata and Chu 2003).

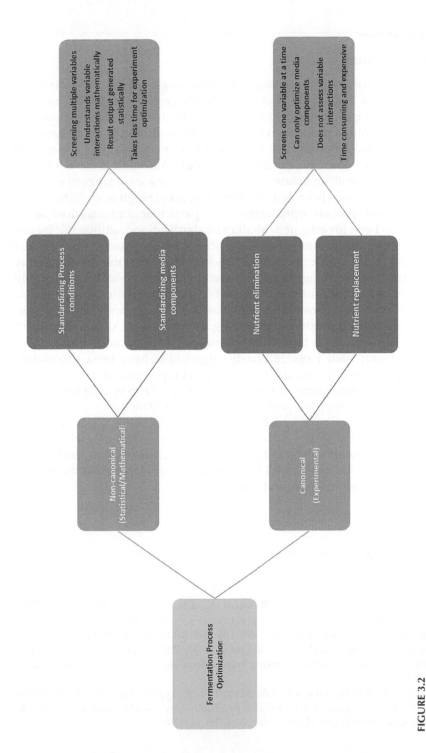

FIGURE 3.2
Strategies for fermentation process optimization.

3.5 Conclusions

The global demand for economically relevant enzymes, that have far-reaching applications in the field of energy, pharmaceutical, food, etc., has been extensively met by fermentation technology. This is due to the long-standing success fermentation processes have had traditionally. But in recent times, the need for sustainable production of enzymes, in a cost-effective eco-friendly manner has burdened the existing technology. Although there are several microbial strains with an innate capacity to produce a wide range of enzymes, the net yield has been below par (Dubey et al. 2008; Singh et al. 2009).

Fermentation process optimization, in particular, optimization of the medium has been investigated. Media standardization traditionally focuses on modulating media components and then tries to assess the most suited component for a said fermentation process. However, the classic method doesn't teach us anything about how individual components interact with each other. Also, the traditional methods have been quite labor-intensive, expensive, and time-consuming. This has led to large-scale investigations on finding alternate means of optimizing the processes. Advances in mathematical/statistical tools have allowed researchers to optimize fermentation processes using such tools. Stochastic algorithms have been successfully developed to not only standardize various components in a fermentation media (such as eliminating or replacing carbon, nitrogen, etc.) but varying process conditions (pH, temperature, etc.) too simultaneously. These simulation tools permit multiple variables to be tested to assess the best product outcome (Wang et al. 2011; Gupte and Kulkarni 2003). Not only this, modern optimization techniques help in reducing the experimental time and the cost of production. In the long run, all of these advantages have made the biotechnology industry eco-friendly and cost-effective.

References

Aftab, M.N., Zafar, A., Iqbal, I., Kaleem, A., Zia, K.M. and Awan, A.R. 2018. Optimization of saccharification potential of recombinant xylanase from *Bacillus licheniformis*. *Bioengineered*. 9(1): 159–165. doi: 10.1080/21655979.2017.1373918.

Ahamad, M.Z., Panda, B.P., Javed, S. and Ali, M. 2006. Production of mevastatin by solid-state fermentation using wheat bran as substrate. *Res. J. Microbiol*. 1: 443–447.

Alexeeva, Y.V., E.P. Ivanova, I.Y. Bakunina, T.N. Zvaygintseva and V.V. Mikhailov, 2002. Optimization of glycosidases production by *Pseudoalteromonas issachenkonii* KMM 3549T. *Lett. Applied Microbiol*. 35: 343–346.

Amanullah, A., Blair, R., Nienow, A.W. and Thomas, C.R. 1999. Effect of agitation intensity on mycelial morphology and protein production in chemostat cultures of recombinant. *Aspergillus oryzae. Biotechnol. Bioeng.* 62: 434–446.

Anderson, R.K.I. and Jayaraman, K. 2005. Impact of balanced substrate flux on the metabolic process employing fuzzy logic during the cultivation of *Bacillus thuringiensis* var. Galleriae. *World J. Microbiol. Biotechnol.* 21: 127–133.

Arbel, A., Rinard, I.H. and Shinnar, R. 1996. Dynamics and control of fluidized catalytic crackers. *Ind Eng Chem Res.* 35: 2215–2233.

Bajpai, P. 1997. Microbial xylanolytic enzyme system. Properties and application. *Adv. Appl. Microbiol.* 43: 141–194.

Banerjee, R. and Bhattachaaryya, B.C. 1993. Evolutionary operation (EVOP) to optimize three- dimensional biological experiments. *Biotechnol. Bioeng.* 41: 67–71.

Bideaux, C., Alfenore, S. and Cameleyre, X. 2006. Minimization of glycerol production during the high-performance fed-batch ethanolic fermentation process. *Appl. Environ. Microb.* 72: 2134–2140.

Campbell-Platt, G. 1994. Fermented foods-a world perspective. *Food Research International.* 27(3): 253–257.

Collins, F. 1992. Cystic fibrosis: molecular biology and therapeutic implications. *Science.* 256: 774–779.

Dubey, K.K., Ray, A. and Behera, B. 2008. Production of demethylated colchicine through microbial transformation and scale-up process development. *Process Biochem.* 43: 251–257. doi: 10.1016/j.procbio.2007.12.002.

Edgington, S. 1993. Nuclease therapeutics in the clinic. *Bio/Technol.* 11: 580–582.

Fogarty, W. 1983. Microbial enzymes and biotechnology. *Appl. Sci.* 4: 117–139.

Frost, G.M. and Moss, D.A. 1987. Production of enzymes by fermentation. In: Rehm, H.J., Reed, G.W. (Eds) *Biotechnology* (Vol 7a), 65–211, VCH Verlagsgessellschaft mbH, Germany.

Gordon, K. 1987. Production of human tissue plasminogen activator in transgenic mouse milk. *Bio. Technol.* 5: 1183–1187.

Gupte, M. and Kulkarni, P. 2003. A study of antifungal antibiotic production by *Thermomonospora* sp MTCC 3340 using full factorial design. *J. Chem. Technol. Biotechnol.* 78: 605–610. doi: 10.1002/jctb.818.

Headon, D.R. and Walsh, G. 1994. The industrial production of enzymes. *Biotechnol. Adv.* 12(4): 635–646. doi: 10.1016/0734-9750(94)90004-3.

Henson, M.A. 2003. Dynamic modeling and control of yeast cell populations in continuous biochemical reactors. *Comp Chem Eng.* 27: 1185–1199.

Howson, S. and Davis, R. 1983. Production of phytase – hydrolysing enzyme by some fungi. *Enz. Microbiol. Technol.* 5: 377–382.

Ikram-ul-Haq, S., Mahmmod, M.J., Zafar S. and Tehmina S. 2006. Triggering glucosidase production in Trichoderma viride UVNG-4 with nutritional and environmental control. *J. App. Sci. Res.* 2: 884–889.

Irc, F.S. 2011. Influence of cultivation conditions on the production of a protease from Aspergillus carbonarius using submerged fermentation. *Afr. J. Food Sci.* 5, 353–365.

Jackson, R.S. 2000. *Wine Science*, Academic Press, USA, 2.

Kar, B., Banerjee, R. and Bhattachaaryya, B.C. 2002. Optimization of physicochemical parameters for gallic acid production by evolutionary operation-factorial design techniques. *Process Biochem.* 37: 1395–1401.

Kennedy, M. and Krouse, D. 1999. Strategies for improving fermentation medium performance: A review. *J. Ind. Microbiol. Biotechnol.* 23: 456–475.

Kumar, C.G. and Takagi, H. 1999. Microbial alkaline protease: From bio industrial viewpoint. *Biotechnol. Adv.* 17: 561–594.

Kumar, D., Jain, V.K., Shanker, G. and Srivastava, A. 2003. Citric acid production by solid state fermentation using sugar cane bagasse. *Process Biochem.* 38: 1731–1738.

Lambert, P. and J. Meers. 1983. The production of industrial enzymes. *Phil. Trans. of the Royal Soc. London*, B300: 63–282.

McNeil, B. and Harvey, L. 1980. *Fermentation, a Practical Approach.* IRL Press, Oxford.

Moffat, A. 1992. High tech plants promise a bumper crop of new products. *Science* 256, 770–771.

Moser, A. and Starzak, M. 1994. Macroapproach kinetics of ethanol fermentation by Saccharomyces cerevisiae. *Chem. Eng. J.* 54: 221–240.

Murat, E. and Dursun, O. 2000. Influence of oxygen transfer on lipase production by *Rhizopus arrhizus. Process Biochem.* 36: 325–329.

Nagata, Y. and Chu, K.H. 2003. Optimization of a fermentation medium using neural networks and genetic algorithms. *Biotechnol. Lett.* 25: 1837–1842.

Nahas, E. and Waldemarin, M.M. 2002. Control of amylase production and growth characteristics of *Aspergillus ochraceus. Revista Latinoamericana de Microbiologia.* 44: 5–10.

Naveena, B.J., Altaf, M., Bhadriah, K. and Reddy, G. 2005. Selection of medium components by Plackett–Burman design for production of L (+) lactic acid by *Lactobacillus amylophilus* GV6 in SSF using wheat bran. *Bioresour. Technol.* 96: 485–490.

Niyonzima, F.C. 2019. Production of Microbial Industrial Enzymes. *ASMI.* 2(12): 75–89. doi: 10.31080/ASMI.2019.02.0434.

Patnaik, P.R. 2005. Neural network designs for poly-b-hydroxybutyrate production optimization under simulated industrial conditions. *Biotechnol. Lett.* 2(7): 409–415.

Pen, J. 1992. Production of active *Bacillus licheniformis* alpha-amylase in tobacco and its application in starch liquification. *Bio. Technol.* 10: 292–296.

Plackett, R.L. and Burman, J.P. 1946. The design of optimum multifactorial experiments. *Biometrika.* 33: 305–325.

Ratzkin, B. 1981. Expression of a gene coding for human urokinase in *E. coli. Proc. of the Nat. Acad. of Sci. USA*, 78: 3313–3317.

Renge, V., Khedkar, S. and Nandurkar, N.R. 2012. Enzyme synthesis by fermentation method: a review. *Sci Rev Chem Comm.* 2(4): 585–590.

Riess, J.G. and LeBlanc, M. 1982. Solubility of and transport phenomena in per-fluorochemicals relevant to blood substitution and other biomedical applications. *Pure Appl. Chem.* 54: 2383–2406.

Rose, A.H. 1980. *Economic Microbiology. Vol. 5, Microbial Enzymes and Bioconversions.* Academic Press, London.

Sayyad, S.A., Panda, B.P., Javed, S. and Ali, M. 2007. Optimization of nutrient parameters for lovastatin production by *Monascus purpureus* MTCC 369 under submerged fermentation using response surface methodology. *Appl. Microbiol. Biot.* 73: 1054–1058.

Schmidt, F.R. 2005. Optimization and scale up of industrial fermentation processes. *Appl. Microbiol. Biotechnol.* 68: 425–435.

Schügerl, K. 2001. Progress in monitoring, modelling and control of bioprocesses during the last 20 years. *J. Biotechnol.* 85: 149–173.

Schwardt, E. 1990. Production and use of enzymes degrading starch and some other polysaccharides. *Food Biotechnol.* 4: 337–351.

Sener, A. and Canbas, A. 2007. The effect of fermentation temperature on the growth kinetics of wine yeast species. *Turk. J. Agric.* 31: 349–354.

Singh, V., Haque, S., Niwas, R., Srivastava, A., Pasupuleti, M. and Tripathi, C.K.M. 2017. Strategies for fermentation medium optimization: An in-depth review. *Front. Microbiol.* 7: 2087. doi: 10.3389/fmicb.2016.02087.

Singh, V., Khan, M., Khan, S. and Tripathi, C.K. 2009. Optimization of actinomycin V production by *Streptomyces triostinicus* using artificial neural network and genetic algorithm. *Appl. Microbiol. Biotechnol.* 82: 379–385. doi: 10.1007/s00253-008-1828-0.

Stanbury, P.F., Whitakar, A. and Hall, S.J. 1997. *Principles of Fermentation Technology.* Elsevier, London, UK.

Starzak M., Krzystek, L., Nowicki, L. and Michalski, H. 1994. Macroapproach kinetics of ethanol fermentation by Saccharomyces cerevisiae. *Chem. Eng. J.* 54: 221–240.

Suganthi, R. 2011. Amylase production by *Aspergillus niger* under solid state fermentation using agroindustrial wastes. *Int. J. Eng. Sci. Technol.* 3(2): 1756–1763.

Svensson, B. and Sogaard, M. 1992. Protein engineering of amylase. *Biochemical Soc. Trans.* 20: 34–42.

Taguchi, H., Yoshida, T., Tomita, Y. and Teramoto, S. 1968. The effects of agitation on disruption of the mycellial pellets in stirred fermenters. *J. Ferment. Technol.* 46: 814–822.

Tunga, R., Banerjee, R. and Bhattachaaryya, B.C. 1999. Optimization of n variable biological experiments by evolutionary operation-factorial design techniques. *J. Biosci. Bioeng.* 87: 125–131.

Ul-Haq, I. and Mukhtar, H. 2006. Fuzzy logic control of bioreactor for enhanced biosynthesis of alkaline protease by an alkalophilic strain of *Bacillus subtilis. Curr. Microbiol.* 52: 224–230.

Wang, Y., Fang, X., An, F., Wang, G. and Zhang, X. 2011. Improvement of antibiotic activity of *Xenorhabdus bovienii* by medium optimization using response surface methodology. *Microb. Cell Fact.* 10: 1–15. doi: 10.1186/1475-2859-10-98.

Wiseman, A. and Gould, B.J. 1971. *Enzymes, Their Nature and Role.* Hutchineson Education Ltd., London, UK.

Yuan, Q.-P., Wang, J.-D., Zhang, H. and Zhang, M.-Q. 2005. Effect of temperature shift on production of xylanase by *Aspergillus niger. Process Biochem.* 40: 3255–3257.

Zhang, J.A., Wei, X.S., Xie, D.M., Sun, Y., Liu, D.H. 2002. Effect of oscillatory temperature and sparging nitrogen gas or carbon dioxide during latter stage of glycerol fermentation. *J. Chem. Ind. Eng.* 53: 980–983.

Zheng, M.Y., Du, G.C., Gu, W.G. and Cheng, J. 2001. A temperature shift strategy in batch microbial transglutaminase fermentation. *Process Biochem.* 36: 525–530.

Zuzek, M., Friedrich, J., Cestnik, B., Karalic, A. and Cimerman, A. 1996. Optimization of fermentation medium by modified method of genetic algorithms. *Biotechnol. Tech.* 10: 991–996.

4

Reforming Process Optimization of Enzyme Production Using Artificial Intelligence and Machine Learning

Rajeev Kumar, Ajay Nair, and Archana S. Rao
Dayananda Sagar University, Bangalore, India

S. M. Veena
Sapthagiri College of Engineering, Bangalore, India

Uday Muddapur
KLE Tach University, Hubli, India

K. S. Anantharaju and Sunil S. More
Dayananda Sagar University, Bangalore, India

CONTENTS

DOI: 10.1201/9781003292333-4

4.1 Introduction

The current understanding of artificial intelligence (AI) can be traced back to the work of McCulloch and Pitts (1990) where they suggested that a suitably designed network could learn (Russel 2010). Alan Turing's (1950) question 'Can machine think?' served as an impetus to the motivation for artificial intelligence. The term 'Artificial Intelligence or AI' was the result of the initiative and pursuance of John McCarthy in the Dartmouth workshop (1956) where his suggestion for this name was adopted. The present interest in Artificial Intelligence is the result of convergence in the field of the processing power of the computer, developments in learning algorithms, and the generation of big data.

AI can be defined as the science and engineering of making intelligent machines. The goals of artificial intelligence are to make computers (machines) with human learning, reasoning, problem-solving ability, and perception capability. The key motivations of this field are to develop computers with the power of human thoughts which can perform behavior similar to humans. AI-driven technology has already started impacting social life. There are two main paradigms of AI research (LeCun et al. 2015) – weak AI and strong AI. Weak AI refers to the AI which involves computers in performing tasks with limited functionality without human cognitive powers. Examples of weak AI include Siri (AI-based voice assistant), Sophia (social humanoid robot), alpha go (a computer program that plays the board game Go) and self-driving cars (driverless cars). Weak AI employs AI methods as a tool for solving problems. Strong AI refers to the generation of computer programs having cognitive states (Searle 1980). Computers with strong AI will possess the ability to think and make decisions like human beings. There is currently no technology with strong AI and is an active area of research.

Machine learning (ML) is the key domain of AI. It is increasingly being used in different industries including manufacturing, advertising, finance, healthcare, and transportation. ML techniques are being used to describe and predict the system using data. These techniques are an important link between data science and AI. Process manufacturing is a highly competitive industrial sector. AI and ML can improve the functioning of manufacturing plants in many ways such as in the improvement of the product quality, identification of bottlenecks in the production processes, automation of different processes, and detection of earlier signs of failure in the equipment.

Enzymes are widely used in many industries – food processing, animal feed, detergent, chemical and pharmaceuticals, biofuel, textile, and leather industry (Leisola 2001). Enzymes are used directly or to make other products. Today, enzymes are used to make over 700 commercial products encompassing over 40 industrial sectors (Patel et al. 2016; Arnau 2020). The current global market for industrial enzymes is about $6.5 billion and is expected to grow at a compound annual growth rate of about 4.9% to become about $8 billion by 2027 (source: PR Newsrwire website). Enzymes are biomolecules that act as a catalyst to support chemical reactions. Enzymes need an optimum temperature, pressure, and pH to function. The activity of enzymes is also sensitive to their substrates or product inhibition. The high cost of isolation, purification, and low yield of enzymes are the common challenges in the production of enzymes (Bhatia 2018; Johannes et al. 2016). The functioning of enzymes under harsh conditions (enzymes used as process catalysts) is one of the major challenges in enzyme production (Saran 2019). The cost of enzyme production is the most important determinant in the production of commercial products (Klein-Marcuschamer et al. 2012). The process optimization of enzymes is a good strategy to reduce the cost of enzyme production (Ferreira et al. 2018).

This chapter is devoted to the applications of AI-ML methods in the process optimization of enzyme production. To appreciate the roles of AI-ML methods a brief overview of different processes involved in enzyme production is given. The need for the process optimizations of key processes is examined. The focus here is on the description of how AI-ML methods are relevant and useful for the process optimization of enzyme production. The various roles of AI-ML methods in process optimizations are explained with some published examples. Finally, the advantages, disadvantages, and future challenges in using AI-ML methods are discussed.

4.2 Process of Enzyme Production

Industrial enzymes are produced from different sources – fungi, bacteria, animals, or plants. The common industrial enzymes from plant sources include phytase, papain, and lipoxygenase. The industrial enzymes derived from animals include trypsin, pepsin, pancreatin, animal lipase, and lysozyme. However, microbes are preferred for the industrial production of enzymes due to ease in the production process. The enzyme production processes using microbes can be divided into the following steps:

1. Isolation of microorganisms: The desirable properties of selecting microorganisms for the isolation process include the high presence of enzyme content, the shorter fermentation time, and the low cost of culture media. Various microbiological techniques are employed to isolate microorganisms from the culture media.

2. Development of strain and preparation of inoculum: The development of suitable strain is the next requirement after the isolation of

the microorganism. The strains are developed using many strain development techniques including mutagens such as ultraviolet lights and chemicals. This step requires the identification of required media composition and culture conditions such as pH and temperature. The media composition is designed to support the growth of the microorganism for the production of enzymes. The other important consideration in formulating media composition is that it should not induce vegetative growth of the organism. A typical medium contains sources of carbon, nitrogen, protein, growth factors, vitamins, and salts. The aerobic and anaerobic nature of the microbes is taken into consideration while formulating the media composition. Care should be taken to maintain the ionic strength of the media (such as the pH of the fermentation medium).

After the development of strain, the inoculum is produced by multiplying its spores or mycelium in the broth or liquid of fermentation. The inoculum is transferred to the fermenter to initiate the fermentation.

3. Fermentation process: The traditional fermentation may be of the following types:

   A. Surface fermentation: In surface fermentation, the inoculum is transferred to the upper surface of the broth.

   B. Submerged fermentation: The inoculum is transferred inside the broth of the fermentation medium in the submerged fermentation medium. (Niyonzima and More 2013). The submerged fermentation is the preferred method of fermentation process due to the lower chance of infection compared to surface fermentation.

   C. Solid-State fermentation: In solid-state fermentation, the cultivation of microorganisms is done on solid substrates: grains, rice, or wheat bran (Lincoln and More 2017)

Based on the mode of operation, the fermentation process can be of two types:

   A. Batch fermentation: In batch fermentation, fermentation is done inside the reactor filled with medium.

   B. Continuous fermentation: In continuous fermentation, fermentation media is continuously added to the reactor and reactor fluid is continuously removed.

4. Enzyme production: During fermentation, growth conditions are maintained. Usually, oil is added to the fermentation medium to suppress the formation of foam. The extracellular enzymes are produced after 30–150 hours of incubation.

5. Isolation of enzyme: Once the fermentation is over, the broth is kept at a low temperature (5°C) to avoid any contamination. Enzymes are isolated from the broth. The isolation process depends on the site of production of enzymes – extracellular or intracellular. This step is easy with fungi where it is done by simple filtration. However, the isolation process is difficult in the case of bacterial culture. In the isolation of enzymes from bacterial culture, calcium salt is added to separate bacterial cells.

6. Purification of enzyme: Enzymes are purified after the isolation step. Any metabolites are removed as metabolites hinder the enzymatic activity. The main steps in purification include the following steps:

   a. The medium containing the enzymes is concentrated using evaporation or ultrafiltration.

   b. The contaminant microbes are removed by filtration or by the addition of chemical compounds such as calcium salts, proteins, starch, sugar, alcohols, sodium chloride sodium, and benzoate.

   c. The enzymes are precipitated using acetone, alcohols, or organic salts.

7. Packaging of enzymes: The purified enzymes are packaged for the commercial supply

The factors influencing enzyme production are called process parameters. The process optimization for enzyme production can be carried out at different levels – laboratory, pilot, and industrial scale. The industrial-scale production range is approximately 1–100 g/L (Kennedy et al. 2009). The industrial productivity of enzymes is about 50 g/L (Patel et al. 2017). A production of 1–100 mg/L at the laboratory is considered good for pilot-scale production (Kennedy et al. 2009).

4.2.1 Need for the Process Optimization in Enzyme Production

The process optimization for enzyme production requires understanding the roles of different factors influencing enzyme production. There are many processes involved (briefly described in Figure 4.1) in the enzyme production process. Typical factors considered in the optimization of enzyme production are strain development medium components, operating conditions (such as temperature and pH), and inoculum (Kennedy and Krouse 1999). Out of these factors, media composition is considered the most important factor. The other factors can become critical in the operation of the production process.

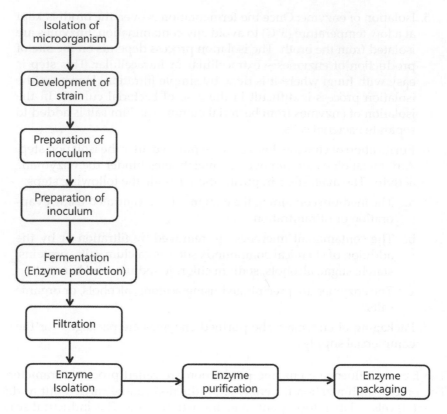

FIGURE 4.1
Steps of enzyme production.

Medium optimization is the most important factor being investigated for process optimization (Singh et al. 2017). The design of media composition is of prime importance because of the following reasons:

1. Media composition affects the product concentration, yield, and volumetric productivity.

2. The nutrient requirements of microorganisms may be different for the different growth phases. The enzyme production is generally high in the stationary phase, but in some cases exponential phase of the growth is important.

3. Medium composition requirement is different for different species. The nutritional requirement may also differ between different strains of the same species.

4. The medium composition can affect the cost of downstream product separation.

5. New strains need to be introduced during the production process which requires that the media composition for the new addition be known.

There are many challenges in designing media. It is not possible to design the media composition in stir-controlled vessels. So, it is done in shake flask systems and the results are extrapolated to the large-scaled stirred tank. However, this extrapolation may not be straightforward (Kennedy et al. 1992) as the two systems are different particularly with pH and oxygen conditions. More data is required to evaluate the scaling up of the media composition based on the data generated from the shake flask systems. Media design in strain development is complex. The media composition is dependent on the strain and the strain development is dependent on the media composition. This complexity needs to be handled during the media design. In practice, the media is taken from published results that may contain expensive components not suitable for industrial production. The published media may also contain too many components whose role may not be known in enzyme production.

There are two broad questions addressed in the media design – Given a set of components in the medium what will be the optimum medium composition for the production of enzyme? (referred to as the closed strategy for the media design) and another question asks – What will be the best combinations of all the possible components available? (referred as the open strategy for the media design). These questions are addressed by performing the experiments. The most plausible experimental design is called factorial design. In a full factorial design, every combination of factor levels is tested. This requires a large number of experimental runs. For example, an experiment with two factors containing seven strains and eight media requires a 7×8 design (56 per replicates) (Thiel et al. 1989). Such an experimental design generates large data which makes the media design a complex process and requires a robust data analysis tool to analyze the data. Machine learning models are equipped to handle such large data for optimizing the media composition. However, a partial factorial experimental design has been proposed and used as an alternative to a full factorial design. Usually, a two-level or three-level (Silveira et al. 1991) factorial design is used to improve the media composition.

Media design is an optimization problem because many components are involved. The relationship between different components and the yield of the enzyme may be simple (linear) or complex (non-linear). The most simple and easy method of the optimization is one-at-a-time approach. In this approach, the concentration of one component is changed keeping everything else constant. This approach provides an easy interpretation of the result, but the optimum is missed if the components show any interaction. The statistical methods are popular in addressing this optimization problem. There are several statistical methods – steepest ascent, canonical, response surface methodology (combination of steepest ascent and canonical method), and Gauss–Siedel method (Kennedy and Krouse 1999). These methods are used as a close strategy and perform well if there is a linear relationship between different components and enzyme production.

The microbial system is a biological system and is subject to dynamic changes during the production processes. Cell metabolism and regulation are important variables in the process optimization of enzyme production. The roles of the cellular system should be considered in the optimization process.

There can be other factors influencing the production of enzymes. These include operating conditions (pH and temperature), airflow rate, stirring flow rate, phases of growth condition of the microbial system (lag phase, exponential, and stationary phase), and filter condition. Any of these factors can be limiting in the production processes of the enzyme.

Therefore, there are a number of factors influencing the production of the enzyme. These factors should be employed in such a way as to keep the cost reasonable so that the production of enzymes remains viable and sustainable. These complexities pose an optimization problem to produce the enzyme.

4.3 Machine Learning Models

Machine learning is a field of computer science that gives computers the ability to learn without being explicitly programmed. It is different from traditional programming where computer programs are written to execute some tasks using the known logic or rules. This is shown in Figure 4.2(A). In machine learning, the output data (expected outcome) is given as the input to the machine learning model, the machine learning model identifies the logic to process the data. Figure 4.2(B) shows the working of the machine learning approach. The difference between traditional programming and machine learning model can be explained by taking an example of the computing area of a circle. In traditional programming, a computer program is written to calculate the area of a circle, i.e., Area = pi * radius * radius (where pi is a constant, the radius is the radius of a circle, and the area is the area of a circle). When an input (value for radius) is given, the program calculates the area of a circle based on the known logic, i.e., the formula for calculating an area of a circle. On the contrary, in the machine learning approach, the computer is supplied with data relating to the area and radius of a circle, and machine learning techniques are used to identify the logic, i.e., a rule for computing the area of a circle. The process by which machine learning methods identify

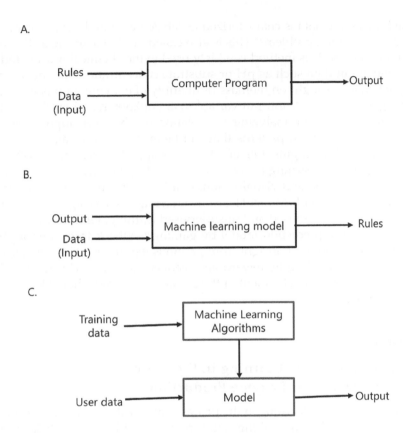

FIGURE 4.2
A: General scheme for traditional programming. B: Scheme for machine learning model. C: Working of a machine learning model.

the logic is called model training. Figure 4.2(C) shows the general scheme of the working of machine learning models.

The inputs to the machine learning models are referred to by different names – input variables, independent variables, predictors, or features. Similarly, the output variables are also referred to by different names – output variables or dependent variables. In the context of enzyme production, the various process parameters are called inputs and the yield of an enzyme is the output.

4.3.1 Types of Machine Learning Model

Machine learning can be of three types – supervised, unsupervised, and reinforcement (Oliver 2017). In supervised learning, machine learning methods are applied in assigning the input data into one of the finite categories known.

In such cases, the data is characterized as labeled data and the problems are called classification problems. This is also called guided learning as the label of data is known. This method is widely used in the chemical process industry where input data such as pH or substrate concentration are known, and the prediction is sought on biomass or quantity of enzyme. In unsupervised learning, the data labels (output variables) are unknown and machine learning methods focus on analyzing the relationship between input variables and uncovering hidden patterns that can be extracted to create new labels regarding possible outputs (Oliver 2017). This approach is known as clustering. Unsupervised learning is very useful with big data where it is used for dimension reduction and visualization. In addition to supervised and unsupervised learning, reinforcement learning is an important category of machine learning. Unlike supervised and unsupervised learning, reinforcement learning is based on a prediction model by gaining feedback from random trial and error and leveraging insight from previous iterations (Oliver 2017). This learning is characterized by reward and feedback learning. This method has been applied in the development of the gaming domain (Bishop 2009).

4.4 Role of Machine Learning in the Process Optimization of Enzyme Production

The roles of machine learning in the process optimization of enzyme production are elucidated by describing two common techniques – artificial neural networks (ANNs) and genetic algorithms.

4.4.1 Artificial Neural Networks (ANN)

The basic idea of ANNs came from the work of Warren McCulloch, a neurophysiologist, and Walter Pitt, a mathematician in 1943 (McCulloch and Pitts 1990). They proposed a neural net based on the 'All or None property' of neural activity. A neuron receives a signal and processes it into binary output. So, a simple McCulloch–Pitt neuron can be used to represent logical operations and is adapted to the machine learning technique.

The ANN consists of nodes that are organized into three types of layers – input, hidden, and output layer. Each layer may contain many neurons. The neuron of one layer is connected to the next layer by a network which is called an edge. Each edge in the network has a numeric weight that can be altered based on experience. If the sum of the connected edges satisfies a set threshold, known as the activation function, this activates a neuron at the next layer. If the sum of the connected edges does not meet the set threshold, the activation function fails, which results in an all-or-nothing arrangement

Input layer   Hidden layer   Output layer

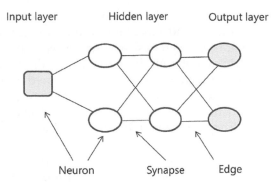

Neuron   Synapse   Edge

FIGURE 4.3
Schematic showing the artificial neural network.

(Oliver 2017). The input and output layer is connected to the hidden layer as shown in Figure 4.3. There can be many hidden layers depending upon the nature of the problem. The connection between two neurons of the different layers is analogous to the biological synapse. A simple schematic of the ANN is depicted in Figure 4.3.

The inputs to the neural networks are called features. Any thing can be used as a feature. For example, in the process optimization of enzyme production, any process parameters can be used as a feature. The process optimization of enzyme production can be set up as a supervised learning problem where enzyme yield or the target variables are output and factors of interest (on which enzyme yield depends) are used as a feature. The neural network model predicted output is compared with the actual output. The difference between the predicted output and actual output is called cost value or cost function. The parameters of the neural network model (e.g., the weight of each edge) are adjusted till the cost function is minimized. Usually, the adjustment is done from the output layer to the input layer. This is called the back-propagating method. The model gets trained when the desired cost value is achieved. This trained model can be used to optimize the process parameters.

This technique is well suited to designing culture media. In designing the culture, the input can be a combination of media components and the enzyme amount can be the output. This formalism can be utilized to find the optimum combination of media components by minimizing the cost function as described above.

In designing the optimal culture media, Imandi and Garapati (2006) used four media components – glutamic acid, citric acid, glycerol, and ammonium chloride, as input and the amount of poly-y-glutamic acid as an output. In their work, the configuration of the neural network was 4-6-1 showing 4 neurons in the input layer, 6 neurons in the hidden layer, and 1 neuron in the output layer. They had used previous data to estimate the weights in the model.

Dutta et al. (2004) used the ANN model to obtain the optimum conditions involving pH, temperature, and inoculum volume for extracellular protease production from a newly synthesized *Pseudomonas* sp. They compared the results of the ANN model with the statistical method response surface method and found more accuracy in ANN model prediction. Ismail et al. (2019) found similar better accuracy for the neural networks model in the optimization of production of exochitinase involving seven variables from *Alternaria* sp. strain. They obtained about 8.5-fold higher production (3.4–28.9 U/dry weight substrate).

Thermoenymes are well suited for harsh industrial processes as they retain their activity under high-temperature conditions. (Khoramnia et al. 2010) employed normal feed-forward back-propagation ANN to identify the optimum operating conditions (pH, temperature, inoculum size, and agitation rpm) for the production of lipase from *Staphylococcushyicus xylosus*. They obtained a 3.5-fold increase in lipase production for the optimum medium. An investigative study (Ebrahimpour et al. 2008) optimized operating conditions (pH, temperature, incubation period, medium volume, inoculum size, and agitation rate) using the ANN model to obtain a 4.7-fold increase in lipase activity from *Geobacillus* sp. strain ARM (DSM 21496 = NCIMB 41583).

The production of microbial enzymes in solid substrate fermentation depends on a number of factors – solid substrates (wheat bran, rice bran, mustard oil cake), incubation temperature, pH of the medium, the particle size of the solid substrate, and moisture level of the substrate. Mishra et al. (2016) studied the effect of these process parameters on the activity of α-amylase activity from *Gliomastix indicus* (MTCC 3869) using the ANN model. They optimized different parametric conditions with mustard oil cake as a substrate using the ANN model to achieve maximum activity of α-amylase activity.

The advantage of this technique is that it is domain-independent and does not require information about the system. Therefore, it is easy to adapt to optimizing the process parameters for enzyme production.

4.4.2 Genetic Algorithms (GA)

The genetic algorithm was developed by J. H. Holland and his collaborator in the 1960s and 1970s. This method was centered on two biological phenomena – natural selection and sexual reproduction. Natural selection determines the survival of an organism based on successful reproduction. Sexual reproduction ensures mixing and recombination among the genes of their offspring (Holland 1975). Thus, the formulation of a genetic algorithm is an abstraction of biological evolution.

A simple GA works in five steps (Melanie 1999). The basic steps in the genetic algorithm are depicted in Figure 4.4. It starts with a randomly

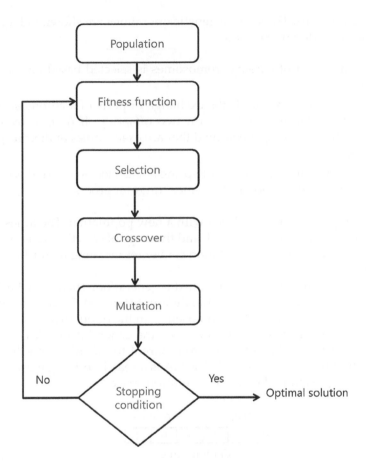

FIGURE 4.4
The steps of genetic algorithms.

generated possible solutions space called 'population'. The possible solution space contains optimal parameter values based on literature or guess (Reddy et al. 1985). The individual of a population represents any possible solution that is likened to a biological chromosome on which reside genes (Haefner 2005). One gene contains a set of parameters. Genes are joined into a string called chromosomes. A chromosome can be represented as bit strings. In the next step, the fitness value is calculated using a fitness function. The fitness function computes a score for each chromosome in the current population. Based on the fitness score, two individuals (parents) are selected. The steps of the genetic operation – selection, crossover, and mutations are performed on selected parents. The steps of genetic operations are shown in Figure 4.4. These steps are described below.

Selection: A pair of parent chromosomes is selected based on the fitness function.

Crossover: The crossover is the exchange of parts of a chromosome at a randomly chosen point to form two offspring. If no crossover takes place, two offspring are formed that are exact copies of their respective parents.

Mutation: Mutation of the two offspring at each locus occurred with some probability which occurs by chance (random process).

The current population is replaced with a new population. The fitness value of the new population is calculated, and the steps of selection, crossover, and mutation are repeated till the solution converges meaning no further change in the fitness value.

Each iteration of this process is called a generation. There may be 50–500 iterations done in a typical run of the different steps. At the end of each run, there is one or more highly fit chromosomes in the population (Figure 4.5).

Thus, the operation of GA steps follows the general principles of natural selection. There is a flow of genes from parent to offspring and the parent with the best fit genes is selected. The GA is a search technique used to find optimal solutions from the available solution space.

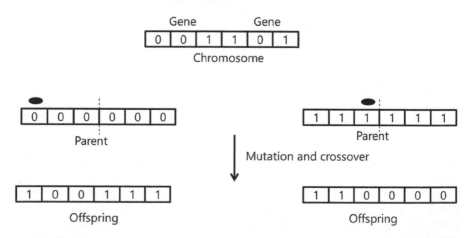

FIGURE 4.5
A chromosome contains genes that are a set of the parameter value. A chromosome is represented as bit strings of two genes. A mutation is a random bit-by-bit change of value. Crossover exchanges portions of bit strings. The dot shows the site of mutation and the dashed line shows the site of crossover.

The GA algorithm is a suitable technique to address the complex problem of media composition optimization for the production of enzymes. Each constituent of media can represent one gene on the chromosome. The genetic operation steps of GA enable the creation of a new combination with a higher fitness value. This new combination can be evaluated for the higher production of enzymes.

The GA can be effectively used to improve the production process of an enzyme. Weuster-Botz and Wandrey (1995) successfully implemented GA in the production process of flavin dehydrogenase (FDH) by the methylotrophic yeast *Candida boldinii*. FDH is produced as an intracellular enzyme by *C. boldinii* using methanol as the energy and carbon source. One of the limiting factors for the production of FDH was the clogging of the filter due to the high salt concentration of nutrient salt concentration. In their work, they optimized the concentration of 14 media components (trace elements and vitamins) within 125 experiments along with the minimized use of nutrient salt concentration. They achieved a 40% improvement in the FDH-specific activity using the optimized media compared with the previously used media.

Mevinolin is a potent competitive inhibitor of hydroxymethylglutaryl-coenzyme A reductase and a cholesterol-lowering agent (Buckland et al. 1989). It is synthesized by the fungus, *Aspergillus terreu*. GA combined with the regression model was used to optimize the fermentation medium (Žužek et al. 1996). They achieved productivity three times compared to the control medium within four generations of fermentation experiments.

GA was used in the optimization of airflow rate (VVM) and stirring rate (RPM) for actinomycin D production from a mutant of *Streptomyces sindenensis*-M-46 (Khan and Tripathi 2011). Using GA, antibiotic concentration was increased by almost 55% as compared to the maximum obtained at the optimum point in the shake flask experiment.

Waste recycling is an important step in reducing environmental pollution. Starch degrading enzymes help in recycling waste. Alternatively, GA was used to optimize three parameters – agitation speed, $\%H_2O_2$, and cultivation time for the production of cellulase from pea hull (Sirohi et al. 2018).

The main advantage of using the GA is to handle a large number of variables (components in media composition). The performance of the ANN depends on the quality of input data. GA can be used to provide a refined set of input data to the neural network model. Nagata and Chu (2003) integrated GA with the ANN model to maximize the production of the enzyme hydantoinase by *Agrobacterium radiobacter*. They developed an ANN model with enzyme concentration as a single output of the model and four medium concentrations as inputs to the model. GA was used to optimize the input space of the neural network model to find the optimal setting for enzyme production. GA combined with the ANN model was used to optimize incubation time, pH, substrate concentration, and inoculum size in the production of L-asparaginase from *Enterobacter aerogenes* MTCC 111 (Reddy et al. 2017). They achieved a 3-fold increase in the enzyme activity using the optimized process parameters.

GA can be modified based on the nature of the problem in the optimization process. The modification involves providing the initial values to the parameter set, range of parameter value (constraint), number of crossover points, and number of generations. A semi-random initial population was used to seed a good population, and remove the worst population, in an attempt to improve the search efficiency in optimizing the eight process variables in the production of laccase by *Daedalea flavida* in stationary culture (Singha and Panda 2014).

In addition to the ANN and GA, the fuzzy logic method is also popular for modeling and optimization of non-linear optimization problems. Fuzzy logic was proposed by Latfi Zadeh and is motivated by human cognition for uncertain decision-making. It is a multivalued logic (between Boolean 0 and 1) used to estimate the membership value of an item to a particular set. Fuzzy logic uses a series of rules using fuzzy membership functions. Ali and Ul Haq (2014) employed fuzzy logic design to improve citric acid production from *Aspergillus niger*. They obtained a several-fold increase in the specific activity of culture. Fuzzy logic can be combined with other machine learning techniques like ANN (Uzuner and Çekmecelioglu 2016) and works particularly well where data is noisy.

4.5 Advantages of Using Artificial Intelligence and Machine Learning

Process optimization of enzyme production requires understanding the effects of various process parameters on enzyme production. Modeling and optimization are important tools to study these effects. Modeling helps in separating different steps involved in the production process and taking important decisions. Identifying the optimum medium composition and operating condition are two important components of process optimization of enzyme production. Media design using full factorial design generates a large amount of data. Machine-learning-based models can reduce the amount of data compared to the full factorial design. Kennedy et al. (1992) investigated three factors (the molasses, ammonium nitrate, and yeast extract concentrations) in the design of a fermentation medium to produce intracellular oil by *Rhodotorula gracilis* and compared the number of experiments required in a full factorial design and neural network model. The factorial design required 27 experiments (each factor at three concentrations) while a neural network model was trained using only 10 experiments for test prediction with reasonable accuracy. Thus, the machine-learning-based approach can help in cost and time reduction in process optimization.

Machine learning can use a large amount of data and discover hidden patterns and associations. In process optimization of enzyme production, a large amount of data is generated at various steps such as media compositions, operating conditions, and functioning of fermenters. Machine learning helps in identifying associations of these data with the amount of enzyme. Machine learning allows using any data as inputs and does not require system-related information (McCord-Nelson and Illingworth 1991). In contrast, mechanistic models require specific inputs and systems structures. The parameters of mechanistic models are difficult to obtain. Machine learning models are free from such constraints. These models are domain independents and hence free from domain expertise. This increases the applicability of the machine learning model for process optimization.

Artificial-intelligence-based optimization model shows better prediction results compared to statistical-regression-based optimization techniques (Haider et al. 2008; Nagata and Chu 2003). The high performance of neural network models is due to their capability of handling non-linear relationships (Lek et al. 1996; Somers and Casal 2009). ANN provides a non-linear mapping between inputs and output (Masoumi et al. 2011). The enzyme production by several microbial systems exhibits an inducible expression pattern (Sahu et al. 2019) which is non-linear in nature. The other popular statistical optimization technique is based on the assumption of linear quadratic correlation for the response (Sahu et al. 2019). Therefore, an ANN has emerged as an attractive tool for the process optimization of enzyme production.

Machine learning models show the property of continuous improvement. This is particularly relevant in the manufacturing industry (Aissani et al. 2008). With more data, models can improve the accuracy and efficiency of prediction.

Machine learning models can be integrated with other machine learning or with mechanistic models. This enables to build hybrid model depending on the nature of the problem. Genetic algorithms can be combined with ANN to identify the optimal inputs. A mechanistic model can be integrated with an ANN to enhance the capability of the overall predictive power of the model. Psichogios and Ungar (1992) had used such a hybrid model for state estimation and prediction in a fed-batch fermenter.

4.6 Disadvantages or Limitations of Machine Learning in the Process Optimization of Enzyme Production

Machine learning models are data-driven models. They require a large data set for training. If the data is not sufficient, they suffer from 'learning difficulty'. The quality of the prediction depends on the quality of the input data. The missing data or noisy data (with large variations) limit the accuracy and efficiency of model prediction results. The overfitting of the model is an important limitation in the application of the machine learning model if the training of the model is overemphasized. In such cases, the model will work well with training and test data but the prediction may not work well with the new data.

Machine learning models are referred to as the black-box approach as they lack a mechanistic basis. This can cause loss of sight from the functioning of the microbial systems. Understanding the functioning of the microbial systems can help in the improvement of the process optimization of enzyme production.

The interpretation of the machine learning model is difficult. The machine learning model is evaluated using different matrices – accuracy, sensitivity, and specificity. There are no set guidelines for selecting a model for the application. Therefore, the use of machine learning in the process optimization of enzyme production requires skilled resources.

4.7 Challenges and Prospects

The application of Machine learning has increased in different fields including manufacturing is due to many factors – availability of data-set, development of computer processing power, and research in ML algorithms development. There has been increased usability and power of available ML tools (Larose 2005). This led to the current increase in interest in machine learning techniques in recent years. However, there are many challenges in using machine learning models (Hoffmann 2014).The most common challenge is the availability of relevant data. The acquisition of data and processing takes a substantial time in setting up the ML model. Sufficient data is not always available for training the model. There are missing values and wide variations in data. Such data is difficult to handle and impacts the performance of the model. Further, selecting the appropriate machine learning technique is a challenge in its application for process optimization. The interpretation of the machine learning model is difficult. The selection of the machine learning model is context-dependent. For example, there should be different criteria for selecting the machine learning model for disease identification and for protein fold prediction. In the former case, a false positive is not desirable but in the latter, false positive is not a constraint in selecting an appropriate model. This consideration underscores the importance of understanding the critical factors in process optimization.

Medium composition optimization is still the key focus area for process optimization as it is the most important factor in the overall cost of enzyme production. The shake flask experiments remain the initial guidance for the design of the media composition for the fermenters. The translation of shake flask data into fermentation medium is not straightforward (Kennedy et al. 1994) and there are not many comparison studies on medium performance have been published. Machine learning can be leveraged in the optimization of medium composition and suitable operating conditions. Most of the published applications of the machine learning model pertain to the label of laboratory or pilot scales. There is a need to apply machine learning models at an industrial scale to assess the success of these methods.

Machine learning models hold great promise for improving the optimization process. A new strategy should be considered for improving process optimization. Understanding the metabolic fluxes and their regulation can be utilized in designing culture medium. Current approaches to medium design treat cells as a constant block and the effect of dynamic cellular regulation is not used (Zhu et al. 1996). Machine learning approaches can help in the optimization of culture medium taking the dynamic regulation into consideration.

Computational fluid dynamics (CFD) models are useful in improving bioreactor configuration for the production of enzymes. These models are used to study the effects of bioreactor design variables on hydrodynamics, biomass growth, and biomass yield. This approach can be helpful in scaling up the production of enzymes (Cappello et al. 2021). Machine learning models can be combined with CFD models to optimize operating conditions. Such an approach was applied in the production of biofuel (del Rio-Chanona et al. 2019).

4.8 Conclusion

Enzyme production is the most important factor for the growth of the enzyme industry. Increasing enzyme production using process optimization is an appropriate and good strategy to make the enzyme industry sustainable and viable. Artificial intelligence and machine learning have ambitious goals to understand the data and convert it into knowledge-based technology to aid in human decisions. In this chapter, we have discussed and presented the various roles of machine learning in the process optimization for the production of enzymes. Machine learning methods have the potential to increase enzyme production using optimizing the medium composition, operating conditions, and some other critical factors. However, other critical factors may be involved in the operation of industrial enzyme production. Machine learning methods can be used to identify these critical factors. The chapter focused on how machine learning methods can be applied to the process optimization for enzyme production and the potential extension of these methods to improve enzyme production.

References

Aissani, Nassima, Bouziane Beldjilali, and Damien Trentesaux 2008. Use of machine learning for continuous improvement of the real time heterarchical manufacturing control system performances. *International Journal of Industrial and Systems Engineering* 3, no. 4: 474–497.

Ali, Sikander, and Ikram Ul Haq 2014. Process optimization of citric acid production from Aspergillus niger using fuzzy logic design. *Pakistan Journal of Botany* 46, no. 3: 1055–1059.

Arnau, José 2020. Strategies and challenges for the development of industrial enzymes using fungal cell factories. In *Grand Challenges in Fungal Biotechnology*, edited by Debbie Yaver, and Carsten M. Hjort, 179–210. Cham: Springer.

Bhatia, Saurabh 2018. *Introduction to Pharmaceutical Biotechnology, Volume 2; Enzymes, Proteins and Bioinformatics*. Bristol, England: IOP Publishing ltd.

Bishop Christopher, M. 2009. *Pattern recognition and machine learning*. New York, NY: Springer.

Buckland, B., K. Gbewonyo, T. Hallada, L. Kaplan, and P. Masurekar 1989. *Novel Microbial Products for Medicine and Agriculture. Society for Industrial Microbiology*, 161–169. Amsterdam, the Netherlands: Elsevier Science Ltd.

Cappello, Vincenzo, Cécile Plais, Christophe Vial, and Frédéric Augier 2021. Scale-up of aerated bioreactors: CFD validation and application to the enzyme production by Trichoderma reesei. *Chemical Engineering Science* 229: 116033.

del Rio-Chanona, Ehecatl Antonio, Jonathan L. Wagner, Haider Ali, Fabio Fiorelli, Dongda Zhang, and Klaus Hellgardt 2019. Deep learning-based surrogate modeling and optimization for microalgal biofuel production and photobioreactor design. *AIChE Journal* 65, no. 3: 915–923.

Dutta, Jayati Ray, Pranab Kumar Dutta, and Rintu Banerjee 2004. Optimization of culture parameters for extracellular protease production from a newly isolated Pseudomonas sp. using response surface and artificial neural network models. *Process Biochemistry* 39, no. 12: 2193–2198.

Ebrahimpour, Afshin, Raja Noor Zaliha Raja Abd Rahman, Diana Hooi Ean Ch'ng, Mahiran Basri, and Abu Bakar Salleh 2008, A modeling study by response surface methodology and artificial neural network on culture parameters optimization for thermostable lipase production from a newly isolated thermophilic Geobacillus sp. strain ARM. *BMC Biotechnology* 8, no. 1: 1–15.

Ferreira, Rafael da Gama, Adriano Rodrigues Azzoni, and Sindelia Freitas 2018. Techno-economic analysis of the industrial production of a low-cost enzyme using E. coli: The case of recombinant β-glucosidase. *Biotechnology for Biofuels* 11, no. 1: 1–13.

Haefner, James W. 2005. *Modeling Biological Systems: Principles and Applications*, 2nd ed. New Delhi, India: Springer.

Haider, M. A., K. Pakshirajan, A. Singh, and S. Chaudhry 2008. Artificial neural network-genetic algorithm approach to optimize media constituents for enhancing lipase production by a soil microorganism. *Applied Biochemistry and Biotechnology* 144: 225–235.

Hoffmann, Achim G. 2014. General limitations on machine learning. In *ECAI, ECAI'90: Proceedings of the 9th European Conference on Artificial Intelligence*, Stockholm Sweden, 345–347. https://www.prnewswire.com/news-releases/industrial-enzymes-market-size-to-reach-usd-8400-4-million-by-2027-at-a-cagr-of-4-9--valuates-reports-301395010.html

Holland, John H. 1975. *Adaptation in Natural and Artificial Systems*. Ann Arbor, MI: The University of Michigan Press.

Imandi, Babu Sarat, and Hanumantha Rao Garapati 2006. Optimization of culture medium for the production of poly-γ-glutamic acid using artificial neural networks and genetic algorithms. *Research Journal of Microbiology* 1: 520–526.

Ismail, Shaymaa A., Ahmed Serwa, Amira Abood, Bahgat Fayed, Siham A. Ismail, and Amal M. Hashem 2019. A study of the use of deep artificial neural network in the optimization of the production of antifungal exochitinase compared with the response surface methodology. *Jordan Journal of Biological Sciences* 12, no. 5: 543–551.

Johannes, Tyler, Michael R. Simurdiak, and Huimin Zhao 2016. Biocatalysis. *Encyclopedia of Chemical Processing* 1: 101–110.

Kennedy, M., and Donal Krouse 1999. Strategies for improving fermentation medium performance: A review. *Journal of Industrial Microbiology and Biotechnology* 23, no. 6: 456–475.

Kennedy, M.J., S.L. Reader, R.J. Davies, D.A. Rhoades, and H.W. Silby 1994. The scale up of mycelial shake flask fermentations: A case study of gamma-linolenic acid production by Mucor hiemalis IRL 51. *Journal of Industrial Microbiology* 13: 212–216.

Kennedy, M. J., S. L. Reader, D. Krouse, and S. Hinkley 2009. Process optimization strategies for biotechnology products: From discovery to production. In *BIOTECHNOLOGY-Volume IV: Fundamentals in Biotechnology*. https://www.eolss.net/ebooklib/sc_cart.aspx?File=E6-58-04-04

Kennedy, Max J., S. G. Prapulla, and M. S. Thakur 1992. Designing fermentation media: A comparison of neural networks to factorial design Biotechnology techniques. *Biotechnology Techniques* 6, no. 4: 293–298.

Khan, M., and C. K. M. Tripathi 2011. Optimization of fermentation parameters for maximization of actinomycin D production. *Journal of Chemical and Pharmaceutical Research* 3: 281–289.

Khoramnia, Anahita, Oi Ming Lai, Afshin Ebrahimpour, Carynn Josue Tanduba, Tan Siow Voon, and Suriati Mukhlis 2010. Thermostable lipase from a newly isolated Staphylococcus xylosus strain; process optimization and characterization using RSM and ANN. *Electronic Journal of Biotechnology* 13, no. 5: 15–16.

Klein-Marcuschamer, Daniel, Piotr Oleskowicz-Popiel, Blake A. Simmons, and Harvey W. Blanch 2012. The challenge of enzyme cost in the production of lignocellulosic biofuels. *Biotechnology and Bioengineering* 109, no. 4: 1083–1087.

Larose, Daniel T. 2005. *Discovering Knowledge in Data – An Introduction to Data Mining*. Hoboken, NJ: Wiley.

LeCun, Yann, Yoshua Bengio, and Geoffrey Hinton 2015. Deep learning. *Nature* 521, no. 7553: 436–444.

Leisola, Matti 2001. *In Industrial Use of Enzymes*, edited by Jouni Jokela, Ossi Pastinen, Ossi Turunen, and Hans Schoemaker. Oxford: Eolss Publishers.

Lek, Sovan, Marc Delacoste, Philippe Baran, Ioannis Dimopoulos, Jacques Lauga, and Stéphane Aulagnier 1996. Application of neural networks to modelling nonlinear relationships in ecology. *Ecological Modelling* 90, no. 1: 39–52.

Lincoln, Lynette, and Sunil S. More 2017. Screening and enhanced production of neutral invertase from Aspergillus sp. by utilization of Molasses–A by-product of Sugarcane industry. *Advances in Bioresearch* 8, no. 4: 103–110.

Masoumi, Hamid Reza Fard, Anuar Kassim, Mahiran Basri, Dzulkifly Kuang Abdullah, and Mohd Jelas Haron 2011. Multivariate optimization in the biosynthesis of a triethanolamine (TEA)-based esterquat cationic surfactant using an artificial neural network. *Molecules* 16, no. 7: 5538–5549.

McCord-Nelson, Marilyn, and William T. Illingworth 1991. *A Practical Guide to Neural Nets*. Boston, MA: Addison-Wesley Longman Publishing Co., Inc.

McCulloch, Warren S., and Walter Pitts 1990. A logical calculus of the ideas immanent in nervous activity. *Bulletin of Mathematical Biology* 52, no. 1: 99–115.

Melanie, Mitchell 1999. *An Introduction to Genetic Algorithms*. Cambridge, MA/ London, England: MIT Press.

Mishra, Santosh Kumar, Shashi Kumar, Surendra Kumar, and Ravi Kant Singh 2016. Optimization of process parameters for-amylase production using Artificial Neural Network (ANN) on agricultural wastes. *Current Trends in Biotechnology and Pharmacy* 10, no. 3: 248–260.

Moor, J. 2006. The Dartmouth College artificial intelligence conference: The next fifty years. *AI Magazine* 27, no 4: 87–89

Nagata, Yuko, and Khim Hoong Chu 2003. Optimization of a fermentation medium using neural networks and genetic algorithms. *Biotechnology Letters* 25, no. 21: 1837–1842.

Niyonzima, Francois N., and Sunil S. More 2013. Screening and optimization of cultural parameters for an alkaline protease production by Aspergillus terreus gr. under submerged fermentation. *International Journal of Pharma and Bio Sciences* 4, no. 1: 1016–1028.

Oliver, Theobald 2017. *Machine Learning for Absolute Beginners. A Plain English Introduction*. Stanford (Calif.): Scatterplot Press.

Patel, Anil Kumar, Reeta Rani Singhania, and Ashok Pandey 2016. Novel enzymatic processes applied to the food industry. *Current Opinion in Food Science* 7: 64–72.

Patel, Anil Kumar, Reeta Rani Singhania, and Ashok Pandey 2017. Production, purification, and application of microbial enzymes. In *Biotechnology of Microbial Enzymes*, edited by Arnold L Demain, Goutam Brahmachari, and Jose L Adrio, 13–41. London: Elsevier.

Psichogios, Dimitris C., and Lyle H. Ungar 1992. A hybrid neural network-first principles approach to process modeling. *AIChE Journal* 38, no. 10: 149–1511.

Reddy, Erva Rajeswara, Rajulapati Satish Babu, Potla Durthi Chandrasai, and Pola Madhuri 2017. Neural network modeling and genetic algorithm optimization strategy for the production of Lasparaginase from Novel Enterobacter sp. *Journal of Pharmaceutical Sciences and Research* 9, no. 2: 124–130.

Reddy, V. R., D. N. Baker, and J. N. Jenkins 1985. Validation of GOSSYM: Part II. Mississippi conditions. *Agricultural Systems* 17, no. 3: 133–154.

Russell, Stuart J. 2010. *Artificial Intelligence 2010. Prentice Hall Series in Artificial Intelligence*. Upper Saddle River, NJ: Prentice Hall.

Sahu, Shraddha, Shailendra Singh Shera, and Rathindra Mohan Banik 2019. Optimization of process parameters for cholesterol oxidase production by Streptomyces Olivaceus MTCC 6820. *The Open Biotechnology Journal* 13: 47–58.

Saran, Saurabh 2019. *A Handbook on High Value Fermentation Products*, vol. 2, edited by Vikash Babu, and Asha Chaubey. Human Welfare: John Wiley & Sons.

Searle, John R. 1980, Minds, brains, and programs. *Behavioral and Brain Sciences* 3, no. 3: 417–424.

Silveira, R.G., T. Kakizono, S. Takemoto, N. Nishio, and S. Nagai 1991. Medium optimization by an orthogonal array design for the growth of Methanosarcina barkeri. *Journal of Fermentation and Bioengineering* 72: 20–25.

Singh, Vineeta, Shafiul Haque, Ram Niwas, Akansha Srivastava, Mukesh Pasupuleti, and C.K.M. Tripathi 2017. Strategies for fermentation medium optimization: An in-depth review. *Frontiers in Microbiology* 7: 2087.

Singha, Siddhartha, and Tapobrata Panda 2014. Improved production of laccase by Daedalea flavida: Consideration of evolutionary process optimization and batch-fed culture. *Bioprocess and Biosystems Engineering* 37, no. 3: 493–503.

Sirohi, Ranjna, Anupama Singh, Ayon Tarafdar, and N. C. Shahi 2018. Application of genetic algorithm in modelling and optimization of cellulase production. *Bioresource Technology* 270: 751–754.

Somers, Mark John, and Jose C. Casal 2009. Using artificial neural networks to model nonlinearity: The case of the job satisfaction—Job performance relationship. *Organizational Research Methods* 12, no. 3: 403–417.

Thiel, Teresa, Judith Bramble, and Selena Rogers 1989. Optimum conditions for growth of cyanobacteria on solid media. *FEMS Microbiology Letters* 61: 27–31.

Turing, Alan 1950. Computing machinery and intelligence. *Mind* LIX, no 236: 433–460.

Uzuner, S., and D. Çekmecelioglu 2016. Comparison of artificial neural networks (ANN) and adaptive neuro-fuzzy inference system (ANFIS) models in simulating polygalacturonase production, *BioResources* 11, no. 4: 8676–8685.

Weuster-Botz, Dirk, and Christian Wandrey 1995. Medium optimization by genetic algorithm for continuous production of formate dehydrogenase. *Process Biochemistry* 30, no. 6: 563–571.

Zhu, Y., A. Rinzema, J. Tramper, and J. Bol 1996. Medium design based on stoichiometric analysis of microbial transglutaminase production by Streptoverticillium mobaraense. *Biotechnology and Bioengineering* 50, no. 3: 291–298.

Žužek, Mateja, Jožica Friedrich, Bojan Cestnik, Arem Karalič, and Aleksa Cimerman 1996. Optimization of fermentation medium by a modified method of genetic algorithms. *Biotechnology Techniques* 10, no. 12: 991–996.

5

Scale-Up Models for Chitinase Production, Enzyme Kinetics, and Optimization

P. V. Atheena

Manipal Institute of Technology, Manipal, India

Keyur Raval

National Institute of Technology Karnataka, Surathkal, India

Ritu Raval

Manipal Institute of Technology, Manipal, India

CONTENTS

5.1 Introduction

Chitin ($C_8H_{13}O_5N$) n is a biopolymer found in abundant amounts after cellulose. It was first isolated from mushrooms by Henri Braconnot in 1811 (Muzzarelli et al. 2012). Structurally, it is very similar to cellulose except for the

DOI: 10.1201/9781003292333-5

acetamide group at the second position instead of the hydroxyl group found on the glucose monomer in cellulose. Chemically, chitin is a linear polysaccharide consisting of β (1-4) linked 2-acetamido 2-deoxy-β-D-glucose (N-acetyl-D-glucosamine GlcNAc (Kuddus and Ahmad 2013) and is found in the extracellular matrix of sponges, nematodes, arthropods, fungi, mollusks, etc. Growth in the global fisheries industry has significantly increased the amount of shell waste in the environment and is a source of pollution. Recycling such wastes cost-effectively is a major challenge (Poria et al. 2021). Chitinases, a group of hydrolytic enzymes, catalyze chitin degradation (Figure 5.1).

Chitinases come under glycoside hydrolase (GH) families 18 and 19. Various bacterial strains have produced chitinases, including *Melghiribacillus thermohalophilus* (Mohamed et al. 2019) and *Paenibacillus* (Poria et al. 2021, Asmani et al. 2020). Upscaling microbial chitinase production is difficult due to its inducible nature, low titer, high cost, and susceptibility to severe environments (Singh et al. 2020). The components of growth media are crucial in the efficient production of chitin-degrading enzymes (Box and Draper 1987).

The myriad number of factors that influence the efficiency of chitin production cannot be comprehended via trial-and-error methods conducted in laboratories. Hence, advanced statistical tools play an important role in optimizing the various factors in enzyme production. The use of such methods in process optimization is collectively called DoE or the design of experiment (Zhou et al. 2010). The basic principles of statistical analysis, such as randomization, replication, duplication, and prediction designs are used to optimize enzyme production. An experimental design aims to reduce the number of steps to extract maximum information from a set of data (Reddy et al. 2003). In this way, the important variables are identified whereas the non-significant ones are eliminated. It also helps to study the interactions between variables

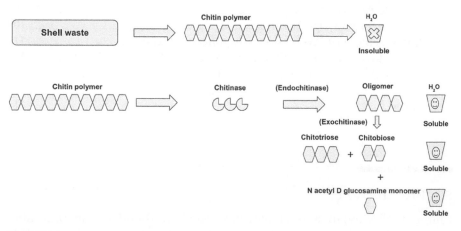

FIGURE 5.1
Mechanism of action of chitinase.

and allows the extrapolation of data. Different levels of significant variables needed for giving the optimum product could also be studied here. Methods in statistics can be broadly divided into univariate experiments and multi-variate experiments. One factor at a time (OFAT) is a univariate approach employed by several researchers for media optimization (Narasimhan et al. 2013a). But it is generally time-consuming and involves many steps (Seltman 2018). Also, the selection of parameters is either random or from previous research. Multivariate experiments are more efficient and considerably reduce the number of experiments to be carried out to optimize a process (Dahiya et al. 2005, Box and Draper 1987). Response surface methodology, process simulation, pattern recognition, etc., are examples of multivariate approaches.

5.2 Response Surface Methodology

Response surface methodology (RSM) is a set of mathematical and statistical techniques which are used to build empirical models. In experiments with a continuous range of values with variables say, x and y, sometimes the true relationship between X and Y is unspecified. The approximation of the response function y = f (x1, x2) + e (where 'e' is the error function) is called response surface methodology (Shen and Morris 2016). They are broadly classified into three groups, namely (a) the first-order, (b) the second-order, and (c) the three-level fractional factorial. If the response can be defined by a linear function of independent variables, then the approximating function is a first-order model. Examples of first-order designs are $2k$ factorial (k is the number of control variables), Plackett–Burman, and simplex designs. In the case of curvature on the response surface, a polynomial with a higher degree should be chosen. Approximating function with two parameters is called a second-order model. Examples are the central composite designs (CCDs) and the Box–Behnken designs (BBDs). In general, all RSM investigations use one of these models or a combination of the two. The RSM's primary purpose is to find the best or optimum response (Bradley 2007). The CCD serves as the base for most of the statistical experiments. The values of the variables are initially optimized with a 'one at a time' approach and later RSM is applied to study the effect of the combination of chosen variables. To produce chitinase, the RSM curve is plotted by keeping chitinase production as a function of other variables. A summary of previous publications that used RSM for optimization was tabulated (Table 5.1). The results obtained are analyzed using analysis of variance (ANOVA) for the interaction of variables and then experimentally verified. As different parameters chosen for screening can have varied units, they are first normalized to have a uniform effect on the dependent variable or the response (Ba and Boyaci 2007) (Figure 5.2).

TABLE 5.1

Summary of Publications with Improvement in Production After Conventional Statistical Optimization

Sl No.	Organism Used	Method of Optimization	Significant Variables After Optimization	Determination Coefficient (R²)	Increase in Production of Chitinase after Optimization	Reference
1)	*Alcaligenes xylosoxydans*	Plackett–Burman Box-Behnken, RSM	Tween20, Yeast extract, chitin	0.889	2.41 folds	Vaidya, Vyas, and Chhatpar (2003)
2)	*Lysinibacillus fusiformis*	Central composite Design, RSM	Chitin, starch, yeast extract, incubations temperature and NaCl	0.569	56.1 folds	R. K. Singh et al. (2013)
3)	*Streptomyces griseorubens*	Plackett–Burman Box-Wilson	Syrup of date, colloidal chitin, yeast extract and K_2HPO_4, KH_2PO_4	0.984	26.38 folds	Meriem and Mahmoud (2017)
4)	*Streptomyces olivaceus*	Box-Behnken	pH, temperature, NaCl concentration, incubation time	0.968	5.09 folds	Sanjivkumar et al. (2020)
5)	*Chitiolyticbacter meiyuanensis*	Plackett–Burman Box-Behnken	Inulin, urea, and sodium sulfate	0.842	15.5 folds	Hao et al. (2012)
6)	*Stenotrophomonas maltophilia*	Plackett–Burman Box-Behnken,RSM	Chitin, maltose, $MgSO_4{\cdot}7H_2O$ KH_2PO_4	0.939	1.06 folds	Khan et al. (2010)
7)	*Paenibacillus sp.*	Plackett–Burman, Box-Behnken,RSM	Urea, K_2HPO_4, $MgSO_4$, yeast extract, and chitin	0.884	2.56 folds	A. K. Singh, Mehta, and Chhatpar (2009)
8)	*Bacillus cereus*	Central composite Design, RSM	Colloidal chitin and KH_2PO_4	0.822	4.4 folds	Kuddus and Roohi (2015)
9)	*Aeromonas schubertii*	Central composite Design, RSM	Temperature and time	0.903	1.54 folds	Liu, Lin, and Juang (2013)
10)	*Bacillus pumilus*	Plackett–Burman, RSM	Chitin concentration, peptone and pH	0.950	6.9 folds	Rishad et al. (2016)

a)Response surface methodology

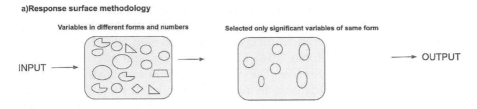

FIGURE 5.2
Mechanism behind the working of response surface methodology.

5.3 Plackett–Burman Design

The Plackett–Burman design is one of the primary tools for the selection of key parameters and the experiment is mostly carried out in a shake flask scale for the same. Plackett–Burman design is appropriate for initial screening even though it does not consider the interactions between independent variables (Vaidya et al. 2003). The important steps involved in Plackett–Burman method can be ordered as below.

a. Identification and selection of factors

b. Attributing the different levels of factors

c. Building the design matrix for Plackett–Burman design

d. Randomizing and performing the experiments

e. Replicating the design

f. Building a model

g. Analyzing the effects

h. Interpretation and conclusion

To start with a screening design, first, the factors must be selected, and their levels have to be defined (Goupy 2005). In an experiment influenced by 50 factors, typically 10–14 factors will be the critical ones. A small set of factors can be efficiently tested by factorial designs (Van Leeuwen et al. 1990). The selection of such factors is largely dependent on the researcher's previous experience and knowledge. Typically, the different variables are tested at two levels (+ maximum, − minimum) in this method. N variables will be screened *via N+1* experiments in Plackett–Burman design. Generally, the response obtained from this model will be in the form of a straight line between two measured variables as the model consider the responses varying linearly with each other (Miller and Sitter 2001). For each factor in the design, new extreme values are attributed to the CCD. These are called star points and there will be twice as many star points as factors in the design (Dilipkumar et al. 2011). To avoid estimation bias in cases where interactions among factors are profound, researchers utilize the 'Augmented Plackett and Burman' designs. The 'foldover' of each run is augmented to eliminate the bias of the main effects (Shen and Morris 2016). Rather than a straight line, the response could also be a curve in this case. The curve can either be inside the measured response results or outside of it. Instead of carrying out the tests in a standard order, they are randomly done to avoid confounding effects (Vanaja and Rani 2007). The idea of a *dummy* variable is also utilized in Plackett–Burman design to help with error estimation efficiently (Singh et al. 2017). After the collection of runs and the calculation of responses, the regression coefficient is estimated. The first-degree polynomial model is utilized in the interpretation of results. After estimating the factor regression coefficients, the ANOVA was used to find significant factors affecting the dependent variables of interest (responses). Key factors identified in Plackett–Burman design are used to devise an experiment using CCD to calculate the response of variables.

5.4 Box–Behnken Experimental Design

Box–Behnken experimental design is another RSM-based tool that is used to optimize process parameters. Most of the time, a Plackett–Burman design is employed to find the key parameters at the first level and BBD is utilized for screening further. Box–Behnken is a three-level complex experimental design. Here, independent variables are studied at three different levels coded ($-1, 0, +1$) indicating low, medium, and high, respectively. BBD is unique in that it does not have combinations in which all components are at their highest or lowest values at the same time. In experiments aiming at understanding the effects of extreme conditions on a variable, BBD is not a wise choice (S. L. C. Ferreira et al. 2007).

The success of the experiment is largely dependent on the parametrization. Process parameters such as process pH, process temperature, inoculation dosages, moisture percentage, particle size, incubation period, media volume, etc., are optimized using this method (Chhaya and Modi 2013). BBD enables (i) the estimation of the quadratic model's parameters, (ii) the creation of sequential designs, (iii) the detection of the model's lack of fit, and (iv) the use of blocks.

5.5 Machine Learning

Machine learning can be defined as the study of algorithms that can improve itself like the human brain improving its accuracy. It is classified under artificial intelligence, which performs by first generating a mathematical model based on sample data fed to the system (training data). The algorithms can make predictions or designs based on the training data without having been explicitly programmed for the task (Jordan and Mitchell 2015). A training set and a test set are presented to the system in machine learning, neural networks, evolutionary computation, and other applications. To create a regressor or classifier, one method in machine learning is to use a training set of data (Goldberg and Holland 1988). The training set will resemble the test set so that it helps in building a model that can predict responses of a test set based on the training set. The two important aspects of machine learning are (a) matrix algebra and (b) optimization. With the currently available data or the sample data, modeling and matrix-vector representation are taken care of by the matrix algebra. This becomes the foundation for a system that can make predictions and decisions on unseen data. Optimization takes care of the numerical parameters required to perform such predictions (Bottou et al. 2018). These machine learning methods can be divided into three subtypes (Buczak and Guven 2016).

a. Based on Network Structure
   Artificial neural network or ANN

b. Based on Statistical Analysis
   Clustering, Hidden Markov Models (HMMs), inductive learning, naive Bayes, etc., come in this category.

c. Based on Evolution
   Genetic algorithms (GAs), genetic programming (GPs), evolution methods, particle swarm optimizations (PSOs) (Kennedy and Eberhart 1995), ant colony optimization (Xu et al. 2010), and artificial immune systems (Farmer et al. 1986) are all examples of evolutionary computation in machine learning.

5.6 Artificial Neural Networks

Our biological nervous system consists of a large number of interconnected processing units called neurons. They communicate with each other with the help of electrical impulses. The human brain is regarded as a massive, highly interconnected network of simple processing units (Abiodun et al. 2018). ANNs are simplified models of the biological nervous system which try to work by initializing the kind of computing performed by human brains. An artificial neuron is also called a perceptron. The input for the neurons first goes to a summation unit. Also, every input will have a specific weight to it.

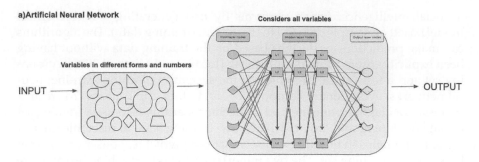

FIGURE 5.3
Mechanism behind the working of artificial neural network.

It is a direct indication of how significant the input is to the neuron. These signals are hence called weighted signals. The input signals and the weights are multiplied and summed in the summation unit before they pass through the threshold unit. The threshold function is also called a transfer function. The output from a threshold function is either '1' or '0', depending on the fact whether the input signal is above or below the threshold value (Dongare et al. 2012). Inside an ANN, there are input layers, hidden layers, and output layers. Each layer delivers its output to the next, and the result is output from the last layer. Between the input and output layers, there exist hidden layers. The output layer determines the final classification category when an ANN is used as a classifier. Although neural networks can take into consideration all the data points, if too much non-significant/less-significant data is fed into the system, the network may exhaust its resources fitting the noise. Therefore, a judicious selection and representation of data are required for the successful implementation of the ANN (Wang et al. 2018; Wang 2003) (Figure 5.3).

5.7 Multilayer Feedforward Networks

A multilayer feedforward network consists of (a) an input layer, (b) hidden layers, and (c) an output layer of neurons. The hidden layers connect the input and output layers and enable the prediction and decision-making from the input provided. The input for neurons in the first hidden layer is from the source nodes of the input layer which supply respective elements of the activation pattern. Like a continuous cycle, now the output from the first hidden layer becomes the input for the second hidden layer. Signals from the last hidden layer become the signals for the output layer. The network's output layer generates a set of neural output signals based on the activation patterns supplied by the source nodes in the network's input layer. This is how the network reacts to the pattern of activation (Zhang et al. 2018). If the number of hidden layers present is more than one, it is then called a deep network. In a feedforward neural network, each neuron takes weights from all the previous layers along with a bias unit. The bias unit is specific to each neuron. In a feedforward network, if we know the X vector (input) along with all the weights (including bias), we can predict the Y vector or the output. In ANN models, there are no assumptions regarding data distribution or properties. As a result, ANNs are more useful for practical applications. ANN models, in contrast to some statistical models that require a hypothesis to be tested, do not require any hypothesis to be tested. Models based on ANNs provide flexible, fault-tolerant data reduction, including non-linear regression and discriminant models. So, they can deal with incomplete data, noise, and non-linear problems. Additionally, trained ANNs can generalize quickly and make accurate predictions.

5.8 Genetic Algorithm

Genetic algorithms are population-based algorithms and comprise computational models inspired by evolution. They are intelligent search techniques maintaining a population of candidate solutions for a given problem and searching the population space by applying various operators. GA mimics the process of natural selection for optimization. In nature, processes like mutation and crossover help in aiding natural selection and thereby the survival of the fittest individuals in a population. A GA tries to replicate this in the best possible manner. To carry out, these GAs have the mutation and crossover operators (Mathew 2012). The other significant parameter here is the search space. A GA design will randomly consider so many different points in a search space aiming to cover the entirety of the search space. The points are spread along the search space to improve the diversity of the population. The random population in a GA is often generated using Gaussian random distribution. GA evaluates the fitness of each individual in the population using a fitness function (Kallel 1998). From each point in the search space with a known fitness function, the best points must be selected. This resembles the selection of fittest individuals from a natural population. This selection results in a scenario where the fittest individuals are contributing more to the production of the next generation. These chosen individuals eventually giving rise to the second generation will have an average fitness higher than that of the first generation due to our selection. This trend continues to further generations forward to obtain the fittest individuals. This is the basic idea of GA. The search in GA is highly exploitative. It is mainly by stochastic operators and not by deterministic rules. By continuously evaluating the fitness function value through generations, one can reach the near optimum. In the actual process, the potential solutions are encoded in a chromosome-like data structure. Encoding can be binary encoding, permutation encoding, real encoding, etc. Parents are selected based on fitness and recombinant operators such as crossover and mutations are applied continuously over generations to preserve the good portions of the chromosome also called a string (Mirjalili 2019). The significant parameters that matter in GA are population size, crossover probability, mutation rate, and mutation type. In general, GA is an easy-to-build and flexible optimization tool that can be used in noisy environments (Whitley 1994) (Figure 5.4).

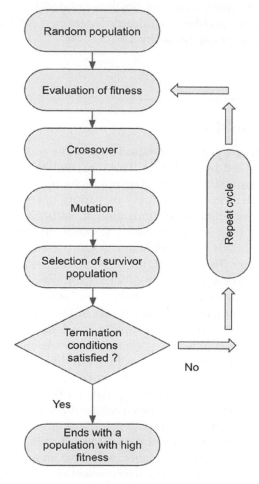

FIGURE 5.4
Mechanism behind the working of genetic algorithm.

5.9 Particle Swarm Optimization

One of the novel approaches that can circumvent conventional optimization problems would be PSO (Suryawanshi and Eswari 2022). Kennedy and Eberhart's first thoughts on particle swarms were primarily aimed at generating computational intelligence by utilizing basic analogs of social interaction. Kennedy and Eberhart first suggested this bio-inspired method in 1995 (Kennedy and Eberhart 1995). PSO starts with several random points on the plane called particles with each particle having two main characteristics, namely position (x) and velocity (v). Each particle will look out for a minimum point in a random direction just the way individual birds in a flock search for their food. The minimum point ever explored by an individual particle as well as the minimum point ever explored by the swarm is considered in arriving at the best value. The individual particle is having no or less power and progress is achieved when interactions of the particle occur (Dongare et al. 2012). PSO has two versions: (a) global and (b) local. For every particle, the best position is called *pbest* and *gbest* for the swarm. In the global version *pbest* and *gbest* are tracked whereas, in the local version, the *pbest* and 'all particles optimal position' called *nbest* is tracked. Several iterations are carried out and *pbest* and *gbest* get updated each time until the purpose is served. The sturdiness of PSO is highly dependent on the swarm size (Poli et al. 2007). An overview of the various elements in the designing of an experiment is shown in Figure 5.5.

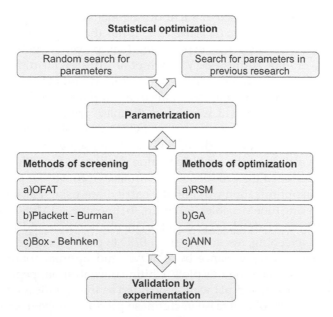

FIGURE 5.5
Overview of design of experiments.

5.10 Enhancement of Production – Significant Studies

A study conducted by Narasimhan et al. on *Bacillus subtilis* for the production of three mycolytic enzymes using OFAT showed significant improvement in chitinase production in an optimized media compared to the unoptimized media. Production was tested for 12 different carbon sources, 8 nitrogen sources, 5 surfactants, and 9 metal ions. The optimized media contained carboxymethyl cellulose (10 g/L), corn steep liquor or KNO_3 (1 g/L), $CaCl_2$, and TritonX100 (0.1% v/v). The optimum pH was found to be 6.0 and 11 for chitin (possibly chitin isomers were present) and the optimum temperature observed was 55°C. The chitinase production was found to increase up to 5.49 folds using this approach (Narasimhan et al. 2013b).

In a study for medium optimization for chitinase production using RSM, interactions of multiple factors like incubation time, initial moisture level, and the ratio of substrate mixture were confirmed (Dahiya et al. 2005). RSM is also applied in finding the values of kinetic constants such as K_m (Michaelis–Menten constant), K_{max} (maximum velocity), E_a (activation energy), temperature coefficient, etc., of a reaction as well. The old method uses the Lineweaver-Burk plot which is highly cumbersome (Beg et al. 2002). In a study by Nawani and Kapadnis, the factors for chitinase production were optimized using RSM in strains of *Streptomyces* and found that it increased by 29% for the strain NK1057, 9.3% for NK528, and 28% for NK951 (Nawani and Kapadnis 2005).

Another similar study used Taguchi orthogonal array for optimization of chitinase production in *Serratia marcescens*. It has the advantage of considering two-factor interactions in an experimental setup even though higher-level interactions were not addressed (V. Singh et al. 2017). Four different factors (NaCl, chitin, pH, temperature) were tested at three different levels to conclude that temperature and pH were more effective compared to other parameters (Zarei et al. 2010). Dahiya et al. worked on chitinase production *via* solid-state fermentation using *Enterobacter* sp. utilized CCD and RSM to improve chitinase production. The model produced 1475 U chitinase per gram of solid substrate which was 97% of the predicted value (Dahiya et al. 2005). A group of researchers optimized the production of chitinase from *Bacillus pumilus* using RSM. Here, the optimization was initially carried out using Plackett–Burman design considering nine variables: (a) chitin, (b) peptone, (c) $MnSO_4$, (d) $FeSO_4$, (e) KH_2PO_4, (f) yeast extract, (g) pH, and (h) incubation time and temperature. After primary screening, researchers used BBD to proceed further and finally, data were analyzed for variance by ANOVA and optimal conditions were identified using surface contour plots. Chitin concentration, peptone, $FeSO_4$, KH_2PO_4, yeast extract, and pH were concluded as the significant parameters. They achieved a 6.9-fold increase in chitinase production from 3.36 U/mL to 23.19 U/mL with the optimization (Beg et al. 2002).

Demir et al. utilized chicken feathers as the chitin source to produce chitinase. Plackett–Burman design was used for parametrization. Individual reactions were carried out in 50 mL Erlenmeyer flasks with shaking. The response of variables could be predicted using a first-order polynomial equation. In 250 mL Erlenmeyer flasks with 50 mL mineral salt medium, samples were transferred and tested at 150 rpm, followed by scale-up in a 2 L stirred tank bioreactor. This model was validated both by Model F value (F) and determination coefficient (R^2). An R^2 value of 0.9356 means the model can explain 93.56% of the variations. Chicken feather which was the chitin source in the experiment was found less significant compared to temperature and time to produce chitinase. The highest chitinase activity obtained was 487 U/mL at 28°C while the predicted model was 412 U/mL (Demir et al. 2015).

Plackett–Burman factorial design and RSM were used to choose the important components of the medium that can affect chitinase production in a study using *Chitinolyticbacter meiyuanensis* (Hao et al. 2012). The study comprised 14 variables in total. For the first time, inulin was used as the primary carbon source in this study. The maximum value of chitinase activity was 15.17 U/mL which was 15-fold that of the original chitinase activity which proves the credibility of the experimental design. These methods are therefore regarded as reliable in picking the vital factors from the multitude of factors affecting the enzyme scale-up hence a compiled list of significant publications was tabulated (Table 5.1).

In a study by Suryawanshi et al., a swarm size of 40 with an inertia weight of 0.2 was studied for 50 generations by applying PSO using MATLAB software to optimize the media components and increase the chitinase production. In the polynomial equation applied to PSO, chitinase activity was taken as the dependent variable Y with colloidal chitin, tween 80, glucose, and yeast extract as independent variables. The model could improve the chitinase activity to 115.8 U/mL (Suryawanshi and Eswari 2022).

Ismail et al. attempted to use deep ANN instead of RSM to improve the exochitinase production from *Alternaria* sp. (Ismail et al. 2019). There was a significant increase of 8.5 folds in the production of exochitinase with a determination coefficient of 0.996 whereas in RSM the R^2 was 0.76. Seven key parameters were screened using Plackett–Burman factorial design. Temperature, the addition of sugarcane bagasse, and the volume of the moistening agent were found to be the significant variables and consequently were subjected to a further step of optimization. BBD was applied to optimize the most significant variables. Since the R^2 value of the model was less than 0.9, ANN has been examined. Deep ANN which is more complex than the normal ANN was employed here. It had three hidden layers (8, 7, and 8 neurons), and the analysis of the data showed an R^2 value of 0.996 which is much better compared to the RSM value. 16.521 U/g was the maximum exochitinase activity achieved with the optimized medium containing (g/flask) wheat bran; 5 g, sugarcane bagasse; 1 g and chitin; 75 mg moistened by 5 mL SR salt solution which was then inoculated by 2 mL of spore suspension of a five-day-old fungus and incubated for sixteen days at 27°C (Ismail et al. 2019). In another study for optimization of process parameters for improved chitinase activity from *Thermomyces* sp. ANN was employed (Suryawanshi et al. 2020). The output layer was the chitinase activity, three neurons in the input layer were incubation time, pH, and inoculum size, and five other neurons were in the hidden layer. The highest value of chitinase production obtained was 128.53 U/mL and confirmed *via* experiment.

5.11 Advantages of Process Optimization Using Advanced Tools

Statistical models such as RSM, OFAT, etc., have the limitation of testing a large number of parameters influencing production at different levels. Methods like ANN and GA can be employed at such times to get more accurate results. Data coverage is more in ANN compared to RSM which in turn reflects in the accuracy of ANN models (Yazdi et al. 2010). ANNs are gaining popularity among researchers recently. The major attraction of ANN is the capability to recognize non-linear and complex parameters and their interactions (Schmidt 2005). In most real-time fermentation systems, the relationship between variables in enzyme production is non-linear. Since ANN models are highly empirical, with sufficient data, ANN models are the best performing, flexible, and easy to fit. Ideally, a feedforward neural network is carried out for the modeling and process optimization and the output is further analyzed by ANN and GA (Subba Rao et al. 2008). The multilayer feedforward neural network consisting of the input layer which is the parameter; the output layer which is the chitinase activity, and hidden nodes that modify the input layer to the benefit of the output layer is used to model chitinase production (Chhaya and Modi 2013). The summation of the variation between the experimental value and the predicted value is calculated as the mean squared error (MSE). A lower MSE is regarded as a good fit model.

5.12 Role of ANN in Scale-Up of Fermentation

Data from the statistical tools are employed in the development of large-scale fermenters to predict the variables that affect production accurately. Discrete signals from the biosensors and the non-linear effect of the variables on production often make it difficult to get control over the fermentation processes.

A study used an ANN that evaluated biosensor measurements to control alcohol fed-batch fermentation (Ferreira et al. 2001). Multilayer neural network was employed for the study. The input layer had two neurons in the net while the output layer had one neuron and the number of neurons was varied from two to five in the hidden layer. Substrate concentration was the controlled variable while the feed flow rate was manipulated. Even though the nets showed a performance that was superior to the experimental data, this could be regarded as a first step taken to control biosensors using ANN in industries. A robust soft sensor to monitor the 1,3-propanediol fermentation process by *Clostridium butyricum* based on an ANN was successfully built by Zhang et al. (2020). Advanced neural networks like the deep ANN can recognize even non-linear and highly complex, non-obvious, and hidden relations between various process parameters and design the process models accordingly. As a result of biosensors and other systems, they allow for instant and short-term responses to process deviations by anticipating physiological drifts and their effects on yield and quality, as well as adjusting the set points and process profiles accordingly (Schmidt 2005). Thus, neural network tools can be used to optimize process control and identify suitable strategies to scale up process optimization. Also, it is a data-driven algorithm that can approximate any reasonable function arbitrarily well. ANN does not need to know the intricate relationship between the variables prior rather it changes its structure based on learning with the help of an interconnected network of artificial neurons (Geyikçi et al. 2012) whereas in RSM the unknown form of relationship between factors and response is approximated. The coefficient of determination (R^2) is used to explain the overall efficiency of a statistical model. A good model should have an R^2 value of less than one. In RSM the addition of variables will increase the value of R^2 irrespective of its statistical significance. This limits the number as well as the form of variables that can be tested in RSM. RSM is mostly used to obtain the maximum or minimum response and their corresponding optimum conditions by considering significant parameters alone (Nair et al. 2014). ANN on the other hand uses all data points to arrive at the response value and is flexible with the number as well as the form of the data (Bezerra et al. 2008). A compiled list of significant publications that compare RSM and ANN for their efficiency in improving production was tabulated (Table 5.2).

5.13 Conclusion

Over the past two decades, technological advancements embracing the areas of machine learning, ANNs, etc., have made transformational progress. These novel technologies have the potential to circumvent the shortcomings

TABLE 5.2

Summary of Publications that Compares Improvement in Production Using RSM and ANN

Sl No.	Aim of the Study	Determination Coefficient (R^2) RSM	Determination Coefficient (R^2) ANN	Reference
1)	Analysis of heavy metal contamination	0.67	0.89	Geyikçi et al. (2012)
2)	Extraction of Oleonolic acid from *Ocimum sanctum*	0.91	0.94	Khamparia et al. (2020)
3)	Compressive strength of recycled concrete aggregates	0.98	0.99	Hammoudi et al. (2019)
4)	Alkylbenzene synthesis over H14[NaP5W30O110]/ SiO2 catalyst	0.92	0.99	Hafizi et al. (2013)
5)	Extraction of artemisinin from *Artemisia annua*	0.90	0.99	Pilkington, Preston, and Gomes (2014)
6)	Modeling of waste coconut oil ethyl esters production	0.95	0.99	Samuel and Okwu (2019)
7)	Fluid extraction of phytochemicals from *Terminalia chebula* pulp	0.997	0.998	Jha and Sit (2021)
8)	Production of bioethanol from pumpkin peel wastes	0.97	0.99	Chouaibi et al. (2020)
9)	Optimization of gypsum-bonded fiberboards	0.90	0.98	Nazerian et al. (2018)
10)	Optimization of Aegle marmelos Oil Extraction for Biodiesel Production	0.97	0.99	Selvan et al. (2018)

of conventional methods used in modeling and optimization. RSM is known to be a popular tool used in optimization and has been utilized by several industries to optimize and enhance their production. It is an excellent system that can optimize production by predicting the response of variables based on a combination of factor levels. Thorough knowledge of the experiment and previous research can be very helpful in maximizing the validity of RSM techniques. ANNs, genetic algorithms, etc., are achieving a lot of attention from researchers and scientists. Prior knowledge about the process is not a prerequisite for such advanced methods as they are more about building connections between all the possible data points to obtain the output. Advanced technologies like PSO are highly promising. The primary goal is to find a fully adapting and parameter-free optimizer. Such technologies involve tuning parameters and selecting significant parameters as they advance from generation to generation. As a result of applying machine knowledge to enzyme production, the cost, number of steps, etc., can be reduced significantly in the future and thereby overall feasibility of scaling up production can be enhanced.

References

Abiodun, Oludare Isaac, Aman Jantan, Abiodun Esther Omolara, Kemi Victoria Dada, Nachaat AbdElatif Mohamed, and Humaira Arshad. 2018. "State-of-the-Art in Artificial Neural Network Applications: A Survey." *Heliyon* 4 (11). Elsevier: e00938.

Asmani, Katia Louiza, Khelifa Bouacem, Akli Ouelhadj, Merzouk Yahiaoui, Sofiane Bechami, Sondes Mechri, Fadoua Jabeur, Kahina Taleb-Ait Menguellet, and Bassem Jaouadi. 2020. "Biochemical and Molecular Characterization of an Acido-Thermostable Endo-Chitinase from *Bacillus Altitudinis* KA15 for Industrial Degradation of Chitinous Waste." *Carbohydrate Research* 495 (September). Elsevier Ltd. doi:10.1016/j.carres.2020.108089.

Ba, Deniz, and Ismail H. Boyaci. 2007. "Modeling and Optimization I: Usability of Response Surface Methodology." *Journal of Food Engineering* 78 (3). Elsevier Ltd: 836–45. doi:10.1016/j.jfoodeng.2005.11.024.

Beg, Qasim Khalil, R. K. Saxena, and Rani Gupta. 2002. "Kinetic Constants Determination for an Alkaline Protease from *Bacillus Mojavensis* Using Response Surface Methodology." *Biotechnology and Bioengineering* 78 (3): 289–95. doi:10.1002/bit.10203.

Bezerra, Marcos Almeida, Ricardo Erthal Santelli, Eliane Padua Oliveira, Leonardo Silveira Villar, and Luciane Amélia Escaleira. 2008. "Response Surface Methodology (RSM) as a Tool for Optimization in Analytical Chemistry." *Talanta*. Elsevier. doi:10.1016/j.talanta.2008.05.019.

Bottou, Léon, Frank E. Curtis, and Jorge Nocedal. 2018. "Optimization Methods for Large-Scale Machine Learning." *Siam Review* 60 (2). SIAM: 223–311.

Box, George E. P., and Norman R. Draper. 1987. *Empirical Model-Building and Response Surfaces*. John Wiley & Sons.

Bradley, Nuran. 2007. "The Response Surface Methodology." PhD diss., Indiana University South Bend.

Buczak, Anna L., and Erhan Guven. 2016. "A Survey of Data Mining and Machine Learning Methods for Cyber Security Intrusion Detection." *IEEE Communications Surveys and Tutorials* 18 (2). Institute of Electrical and Electronics Engineers Inc.: 1153–76. doi:10.1109/COMST.2015.2494502.

Chhaya, Ronak, and H. A. Modi. 2013. "Submerged Fermentation of Laccase Producing *Streptomyces Chartreusis* Using Box-Behnken Experimental Design." *Journal of Pharmacy and Biological Sciences* 6 (4): 46–53.

Dahiya, Neetu, Rupinder Tewari, Ram Prakash Tiwari, and Gurinder Singh Hoondal. 2005. "Chitinase Production in Solid-State Fermentation by *Enterobacter Sp.* NRG4 Using Statistical Experimental Design." *Current Microbiology* 51 (4): 222–28. doi:10.1007/s00284-005-4520-y.

Demir, Tuğçe, E. Esin Hameş, Suphi S. Öncel, and Fazilet Vardar-Sukan. 2015. "An Optimization Approach to Scale up Keratinase Production By *Streptomyces Sp.* 2M21 by Utilizing Chicken Feather." *International Biodeterioration and Biodegradation* 103 (September). Elsevier Ltd: 134–40. doi:10.1016/j.ibiod.2015.04.025.

Dilipkumar, M., M. Rajasimman, and N. Rajamohan. 2011. "Optimization of Inulinase Production from Garlic by *Streptomyces Sp.* in Solid State Fermentation Using Statistical Designs." *Biotechnology Research International* 2011 (March). Hindawi Limited: 1–7. doi:10.4061/2011/708043.

Dongare, A. D., R. R. Kharde, Amit D. Kachare, et al. 2012. "Introduction to Artificial Neural Network." *International Journal of Engineering and Innovative Technology (IJEIT)* 2 (1). Citeseer: 189–94.

Farmer, J. Doyne, Norman H. Packard, and Alan S. Perelson. 1986. "The immune system, adaptation, and machine learning." *Physica D: Nonlinear Phenomena* 22. Elsevier: 187–204.

Ferreira, L. S., M. B. De Souza Jr, and R. O. M. Folly. 2001. "Development of an Alcohol Fermentation Control System Based on Biosensor Measurements Interpreted by Neural Networks." *Sensors and Actuators B: Chemical* 75 (3). Elsevier: 166–71.

Ferreira, S. L.C., R. E. Bruns, H. S. Ferreira, G. D. Matos, J. M. David, G. C. Brandão, E. G.P. da Silva, et al. 2007. "Box-Behnken Design: An Alternative for the Optimization of Analytical Methods." *Analytica Chimica Acta* 597. Elsevier: 179–186. doi:10.1016/j.aca.2007.07.011.

Geyikçi, Feza, Erdal Kiliç, Semra Çoruh, and Sermin Elevli. 2012. "Modelling of Lead Adsorption from Industrial Sludge Leachate on Red Mud by Using RSM and ANN." *Chemical Engineering Journal* 183. Elsevier: 53–59. doi:10.1016/j.cej.2011.12.019.

Goldberg, David E., and John Henry Holland. 1988. "Genetic Algorithms and Machine Learning." In *Machine Learning*, Springer.

Goupy, Jacques. 2005. "What Kind of Experimental Design for Finding and Checking Robustness of Analytical Methods." *Analytica Chimica Acta*, 544. Elsevier: 184–90. doi:10.1016/j.aca.2005.01.051.

Hao, Zhikui, Yujie Cai, Xiangru Liao, Xiaoli Zhang, Zhiyou Fang, and Dabing Zhang. 2012. "Optimization of Nutrition Factors on Chitinase Production from a Newly Isolated *Chitiolyticbacter Meiyuanensis* SYBC-H1." *Brazilian Journal of Microbiology*, 43. Springer: 177–86. http://www.ncbi.nlm.nih.gov.

Ismail, Shaymaa A, Ahmed Serwa, Amira Abood, Bahgat Fayed, Siham A. Ismail, and Amal M. Hashem. 2019. "A Study of the Use of Deep Artificial Neural Network in the Optimization of the Production of Antifungal Exochitinase Compared with the Response Surface Methodology." *Jordan Journal of Biological Sciences* 12 (5).

Jordan, Michael I., and Tom M. Mitchell. 2015. "Machine Learning: Trends, Perspectives, and Prospects." *Science* 349 (6245). American Association for the Advancement of Science: 255–60.

Kallel, Leila. 1998. "Inside GA Dynamics: Ground Basis for Comparison." In *International Conference on Parallel Problem Solving from Nature*, pp. 57–66. Springer, Berlin, Heidelberg.

Kennedy, J., and R. Eberhart. 1995. "Particle Swarm Optimization." In *Proceedings of ICNN'95 - International Conference on Neural Networks*, 4, pp. 1942–48. IEEE. doi:10.1109/ICNN.1995.488968.

Kuddus, M., and I. Z. Ahmad. 2013. "Isolation of Novel Chitinolytic Bacteria and Production Optimization of Extracellular Chitinase." *Journal of Genetic Engineering and Biotechnology* 11 (1). Elsevier: 39–46.

Mathew, Tom V. 2012. "Genetic Algorithm." *Report Submitted at IIT Bombay.*

Miller, A., and R. R. Sitter. 2001. "Using the Folded-over 12-Run Plackett-Burman Design to Consider Interactions." *Technometrics* 43 (1). American Statistical Assoc: 44–55. doi:10.1198/00401700152404318.

Mirjalili, Seyedali. 2019. "Genetic Algorithm." In *Studies in Computational Intelligence*, 780, pp. 43–55. Springer Verlag. doi:10.1007/978-3-319-93025-1_4.

Mohamed, Sara, Khelifa Bouacem, Sondes Mechri, Nariman Ammara Addou, Hassiba Laribi-Habchi, Marie Laure Fardeau, Bassem Jaouadi, Amel Bouanane-Darenfed, and Hocine Hacène. 2019. "Purification and Biochemical Characterization of a Novel Acido-Halotolerant and Thermostable Endochitinase from *Melghiribacillus Thermohalophilus* Strain Nari2AT." *Carbohydrate Research* 473. Elsevier: 46–56. doi:10.1016/j.carres.2018.12.017.

Muzzarelli, Riccardo A A, Joseph Boudrant, Diederick Meyer, Nicola Manno, Marta DeMarchis, and Maurizio G. Paoletti. 2012. "Current Views on Fungal Chitin/Chitosan, Human Chitinases, Food Preservation, Glucans, Pectins and Inulin: A Tribute to Henri Braconnot, Precursor of the Carbohydrate Polymers Science, on the Chitin Bicentennial." *Carbohydrate Polymers* 87 (2). Elsevier: 995–1012.

Nair, Abhilash T., Abhipsa R. Makwana, and M. Mansoor Ahammed. 2014. "The Use of Response Surface Methodology for Modelling and Analysis of Water and Wastewater Treatment Processes: A Review." *Water Science and Technology* 69 (3). IWA Publishing: 464–78. doi:10.2166/wst.2013.733.

Narasimhan, Ashwini, Deepak Bist, Samantha Suresh, and Srividya Shivakumar. 2013a. "Optimization of Mycolytic Enzymes (Chitinase, 1, 3-Glucanase and Cellulase) Production by *Bacillus Subtilis*, a Potential Biocontrol Agent Using One-Factor Approach." NISCAIR-CSIR, India.

——— 2013b. "Optimization of Mycolytic Enzymes (Chitinase, B1,3-Glucanase and Cellulase) Production by *Bacillus Subtilis*, a Potential Biocontrol Agent Using One-Factor Approach." *Journal of Scientific & Industrial Research* 72.

Nawani, N. N., and B. P. Kapadnis. 2005. "Optimization of Chitinase Production Using Statistics Based Experimental Designs." *Process Biochemistry* 40. Elseveir: 651–60. doi:10.1016/j.procbio.2004.01.048.

Poli, Riccardo, James Kennedy, and Tim Blackwell. 2007. "Particle Swarm Optimization: An Overview." *Swarm Intelligence* 1. Springer: 33–57. doi:10.1007/s11721-007-0002-0.

Poria, Vikram, Anuj Rana, Arti Kumari, Jasneet Grewal, Kumar Pranaw, and Surender Singh. 2021. "Current Perspectives on Chitinolytic Enzymes and Their Agro-Industrial Applications." *Biology* 10. MDPI: 1319. doi:10.3390/biology10121319.

Reddy, P. Rama Mohan, B. Ramesh, S. Mrudula, Gopal Reddy, and G. Seenayya. 2003. "Production of Thermostable β-Amylase by *Clostridium Thermosulfurogenes* SV2 in Solid-State Fermentation: Optimization of Nutrient Levels Using Response Surface Methodology." *Process Biochemistry* 39 (3). Elsevier: 267–77.

Schmidt, F. R. 2005. "Optimization and Scale up of Industrial Fermentation Processes." In *Applied Microbiology and Biotechnology*. Springer Verlag. doi:10.1007/s00253-005-0003-0.

Seltman, Howard J. 2018. "Within-Subjects Designs." In *Experimental Design and Analysis*, pp. 339–56. Carnegie Mellon University, Pittsburgh, PA.

Shen, Lu, and Max D. Morris. 2016. "Augmented Plackett–Burman Designs with Replication and Improved Bias Properties." *Journal of Statistical Planning and Inference* 179 (December). Elsevier B.V.: 15–21. doi:10.1016/j.jspi.2016.07.003.

Singh, Raj, Sushil Kumar Upadhyay, Manoj Singh, Indu Sharma, Pooja Sharma, Pooja Kamboj, Adesh Saini, et al. 2020. "Chitin, Chitinases and Chitin Derivatives in Biopharmaceutical, Agricultural and Environmental Perspective." *Biointerface Research in Applied Chemistry*. AMG Transcend Association. doi:10.33263/BRIAC113.998510005.

Singh, Vineeta, Shafiul Haque, Ram Niwas, Akansha Srivastava, Mukesh Pasupuleti, and C. K.M. Tripathi. 2017. "Strategies for Fermentation Medium Optimization: An in-Depth Review." *Frontiers in Microbiology*. Frontiers Research Foundation. doi:10.3389/fmicb.2016.02087.

Subba Rao, Ch, T. Sathish, M. Mahalaxmi, G. Suvarna Laxmi, R. Sreenivas Rao, and R. S. Prakasham. 2008. "Modelling and Optimization of Fermentation Factors for Enhancement of Alkaline Protease Production by Isolated *Bacillus Circulans* Using Feed-Forward Neural Network and Genetic Algorithm." *Journal of Applied Microbiology* 104. John Wiley and Sons: 889–98. doi:10.1111/j.1365-2672.2007.03605.x.

Suryawanshi, Nisha, and J. Satya Eswari. 2022. "Chitin from Seafood Waste: Particle Swarm Optimization and Neural Network Study for the Improved Chitinase Production." *Journal of Chemical Technology and Biotechnology* 97 (2). John Wiley and Sons: 509–19. doi:10.1002/jctb.6656.

Suryawanshi, Nisha, Jyoti Sahu, Yash Moda, and J. Satya Eswari. 2020. "Optimization of Process Parameters for Improved Chitinase Activity from *Thermomyces Sp.* by Using Artificial Neural Network and Genetic Algorithm." *Preparative Biochemistry and Biotechnology* 50 (10). Bellwether Publishing, Ltd.: 1031–41. doi:10.1080/10826068.2020.1780612.

Vaidya, Rajiv, Pranav Vyas, and H. S. Chhatpar. 2003. "Statistical Optimization of Medium Components for the Production of Chitinase by *Alcaligenes Xylosoxydans*." *Enzyme and Microbial Technology* 33 (1). Elsevier Inc.: 92–96. doi:10.1016/S0141-0229(03)00100-5.

Van Leeuwen, J. A., B. G. M. Vandeginste, G. Kateman, M. Mulholland, and A. Cleland. 1990. "An Expert System for the Choice of Factors for a Ruggedness Test in Liquid Chromatography." *Analyrica Chimica Acta*. 228. Elsevier Science Publishers B.V.

Vanaja, K., and R. H. Shobha Rani. 2007. "Design of Experiments: Concept and Applications of Plackett Burman Design." *Clinical Research and Regulatory Affairs* 24 (1). Taylor & Francis: 1–23.

Wang, Dongshu, Dapei Tan, and Lei Liu. 2018. "Particle Swarm Optimization Algorithm: An Overview." *Soft Computing* 22 (2). Springer Verlag: 387–408. doi:10.1007/s00500-016-2474-6.

Wang, Sun-Chong. 2003. "Artificial Neural Network." In *Interdisciplinary Computing in Java Programming*.

Whitley, Darrell. 1994. "A Genetic Algorithm Tutorial." *Statistics and Computing* 4. Springer: 65–85.

Xu, B., J. Zhu, and Q. Chen. 2010. *Ant Colony Optimization. New Advances in Machine Learning*, pp. 1155–73. InTech, Jiangsu, China.

Yazdi, M. R. Soleymani, and A. Khorram. 2010. "Modeling and Optimization of Milling Processby Using RSM and ANN Methods." *International Journal of Engineering and Technology* 2 (5). IACSIT Press: 474.

Zarei, Mandana, Saeed Aminzadeh, Hossein Zolgharnein, Alireza Safahieh, Ahmad Ghoroghi, Abbasali Motallebi, Morteza Daliri, and Abbas Sahebghadam Lotfi. 2010. "*Serratia Marcescens* B4A Chitinase Product Optimization Using Taguchi Approach." *Iranian Journal of Biotechnology* 8 (4). National Institute of Genetic Engineering and Biotechnology: 252–62.

Zhang, Ai Hui, Kai Yi Zhu, Xiao Yan Zhuang, Lang Xing Liao, Shi Yang Huang, Chuan Yi Yao, and Bai Shan Fang. 2020. "A Robust Soft Sensor to Monitor 1,3-Propanediol Fermentation Process by *Clostridium Butyricum* Based on Artificial Neural Network." *Biotechnology and Bioengineering* 117. John Wiley and Sons: 3345–55. doi:10.1002/bit.27507.

Zhang, Zhihua, Khelifi Zhang, and A Khelifi. 2018. *Multivariate Time Series Analysis in Climate and Environmental Research*. Springer, Cham.

Zhou, Wen Wen, Yun Long He, Tian Gui Niu, and Jian Jiang Zhong. 2010. "Optimization of Fermentation Conditions for Production of Anti-TMV Extracellular Ribonuclease by *Bacillus Cereus* Using Response Surface Methodology." *Bioprocess and Biosystems Engineering* 33. John Wiley and Sons: 657–63. doi:10.1007/s00449-009-0330-0.

6

Genetic Algorithm for Optimization of Fermentation Processes of Various Enzyme Productions

Karan Kumar

Indian Institute of Technology, Guwahati, India

Heena Shah

Mandsaur University, Mandsaur, India

Vijayanand S. Moholkar

Indian Institute of Technology, Guwahati, India

CONTENTS

DOI: 10.1201/9781003292333-6

6.1 Introduction

Enzymes are the need of the hour and are being recognized globally because of their vast applications in various industrial processes, e.g., agriculture and food processing, medicine, chemical, and green energy production (Malani et al. 2019; Kumar et al. 2021). Enzymes are highly efficient bio-catalysts made up of large bio-macromolecules composed of amino acids (Al-Maqtari et al. 2019). Enzyme-mediated processes are winning the competition against other catalytic processes due to their cost-effectiveness, less processing time, low energy intake, non-toxic, and eco-friendly nature (Lee and Ra 2021). The use of enzymes in such processes is inevitable because of their aforementioned intrinsic properties. Moreover, the global market for industrial enzymes has been valued to reach nearly $6.6 billion in 2021 and is predicted to reach $9.1 billion by 2026 with a compound annual growth rate (CAGR) of 6.6% (Industrial Enzymes Market Global Outlook, Trends, and Forecast to 2026 | MarketsandMarkets 2022). Therefore, interest towards commercial enzyme production has grown due to the expanding market demand for biocatalysts. Moreover, the demand for non-renewable fossil resources is increasing with the accelerated global population; hence, the quest for sustainable and eco-friendly resources is being popularized globally (Malani et al. 2019; Kumar et al. 2021; Singh et al. 2022). Industrial production of enzymes has been achieved via microbial fermentation. Fermentation is a biotechnological process whereby renewable substrates are converted into value-added products such as organic acids, bio-alcohols, therapeutic and commercial enzymes, biopolymers, etc., with the help of microorganisms such as bacteria, yeast, and fungi (Al-Maqtari et al. 2019). Mainly, submerged fermentation (SmF) and solid-state fermentation (SSF) are the two modes that have been widely explored for industrial enzyme production (Erva et al. 2017). Fermentation being a complex process involves optimization of several parameters (e.g., concentration of carbon and nitrogen sources, inoculum size, pH, temperature, fermentation period, aeration rate, etc.) parallelly to obtain efficient enzyme yields. Conventional optimization techniques that deal with "single factor" or one-factor-at-a-time (OFAT), make the process extensive, time-consuming, and expensive with minimal accuracy (Schmidt 2005; Carboué et al. 2017). Considering the serious need for fermentation processes in chemical industries, pharmaceuticals, energy production, etc., a method for the optimization of the fermentation process that will solve the problems with high efficiency and accuracy is a necessity. Mathematical optimization techniques such as genetic algorithms (GAs) can be used to run parallel many process parameters for solving fermentation problems since it uses mathematical modelling systems (Balen et al. 2015; Almadhoun and Hamdan 2018; Mondal et al. 2021). The in-silico process is initialized by building virtual response surface methodology (RSM) or artificial neural network (ANN) models to optimize the medium inputs and then the genetic algorithm is used to obtain optimal values for the input parameters based on the ANN modelling (Roeva 2006; Almadhoun and Hamdan 2018). Compared with conventional optimization methods, GA adds on several points resulting in a better method for optimization of fermentation process as they are the profitable global optimization solution. GA is based on the principles of Darwin's evolutionary theory that deals with the idea of genetics and the process of natural selection, where the historical information is used to take the algorithm to the best level (Holland 1992; Roeva 2006; Zheng et al. 2017; Salim et al. 2019). GA begins with a set of individuals referred to as population, and then offspring produced by the parent are used to create a new population. The selection is done in such a way that the new population should be more suitable compared to the old one (Mondal et al. 2021; Sales de Menezes et al. 2021). This chapter aims to develop a clear understanding of Genetic Algorithms regarding the optimization of fermentation processes. In subsequent sections, we will talk about the process of industrial enzyme production, an overview of conventional and contemporary process optimization techniques followed by the application of state-of-the-art techniques such as GA, ANN and hybrid GA-ANN. Lastly, this chapter concludes with how GA benefits and solves the problems associated with the optimization of process parameters for various fermentative enzyme productions. Moreover, we have discussed the present and future scope of GA in fermentation technology.

6.2 Industrial Production of Enzymes from Microbial Sources

Fermentation is a process in which complex organic molecules are broken down by the action of microorganisms or their enzymes in anaerobic or aerobic conditions. Microorganisms have been widely used as cell factories because they are capable of transforming numerous compounds into value-added products such as organic acids, alcohols, vitamins, enzymes, proteins, and so on. Microorganisms serve as an excellent source of various therapeutic as well as commercial enzymes due to their rapid growth and broad diversity of biochemical nature (Stanbury et al. 2013). The production of various enzymes using fermentation technology has been extensively studied for decades such as cellulases, xylanases, proteases, amylases, lipases, L-glutaminases, β-glucosidases, L-asparaginase, and α-galactosidase. Fermentation is the most widely used technology at an industrial scale due to its economic and environmental advantages (Soccol et al. 2017).

The fermentation process involves a series of operations in the fermentative production of microbial products. Industrial enzyme production via fermentation involves three major stages, viz., upstream, midstream, and downstream. The initial step of fermentation is the upstream processing that entails the media formulation, sterilization, isolation/acquisition, improvement, and cultivation of industrial strains to produce a higher quantity of the desired product(s). Next, the microbial strain is inoculated in a seed tank containing a suitable culture medium for obtaining inoculum or stock culture. After establishing the inoculum, a growth medium has to be formulated which contains all the basic and specific growth requirements of the microorganisms and favours high product concentration, high rate of product turnover and biomass yield. The upstream process is completed with the transfer of sterilized fermentation medium and inoculum in the fermenter to initiate the fermentation process (Mustefa Beyan et al. 2021). Next is the midstream stage, also known as the fermentation stage, this step aims to provide and maintain all the optimum physical and chemical growth parameters to carry out the fermentation process efficiently so that maximum production could be achieved. Because enzymes are highly sensitive to pH and temperature which can lead to their denaturation, it is crucial to maintain all parameters at their optimum levels throughout the fermentation process. (Erva et al. 2017). Therefore, it is essential to know the stoichiometry and energetics of the conversion of substrate into product, kinetics of growth and product turnover, heat and mass transfer rates, etc. The downstream stage includes isolation, extraction, and purification of the product from the fermented broth. The microbial cell synthesizes both extracellular and intracellular enzymes. The recovery of extracellular enzymes is achieved through filtration, while separation of intracellular enzymes needs cell disruption or cell lysis. The separation of enzymes from the spent medium is usually done by centrifugation and filtration followed by product concentration and purification using ultracentrifugation, precipitation, and chromatographic techniques such as ion exchange, gel filtration, etc. (Kiruthika and Murugesan 2020; Rangel et al. 2020). Figure 6.1 is the graphical representation of the steps involved in industrial-scale fermentation process for enzyme production.

6.2.1 Fermentation Methods

The production of enzymes by fermentation is a particularly promising method. Microbial production of different enzymes is mainly carried out using submerged and SSF techniques.

6.2.1.1 Solid-State Fermentation

SSF is a process in which microorganisms grow on solid support in an environment in the absence of water or near absence of water. However, the substrate must possess enough moisture to facilitate the growth and metabolism of microorganisms. SSF resembles the natural habitat of microorganisms.

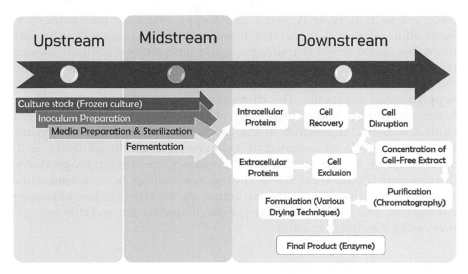

FIGURE 6.1
Schematic representation of steps involved in industrial enzyme production.

The majority of SSF processes involve filamentous fungi and mould (Soccol et al. 2017). In SSF, bacteria and fungi are allowed to grow on a solid matrix or substrates including agriculture residues such as cereal brans, sugarcane bagasse, cassava bagasse, fruit peels and pulps, straws, corn cobs, and paper pulps. These materials are rich in cellulose, hemicelluloses, lignin, and starch. The solid matrix provides two main functions; they serve as a solid support for nutrient absorption and microbial growth, and also as a source of organic nutrients. For optimum growth, the medium is supplemented with essential nutrients. Phosphorus, sulphur, potassium, magnesium, calcium, zinc, manganese, copper, iron, cobalt, and iodine are some of the macro- and micronutrients that are commonly supplied to the medium. The aim of SSF is to bring cultured microorganisms into close proximity with the substrate to achieve and utilize the maximum concentration of nutrients from the substrate for fermentation (Singhania et al. 2009; Soccol et al. 2017; Sadh et al. 2018). SSF is considered a promising method for the production of commercial enzymes. The success of any bioprocess technology depends on various important physico-chemical parameters that have a significant impact on the fermentation process and must be considered while developing bioprocesses, including the SSF.

6.2.1.2 Submerged Fermentation

SmF is a well-established method of cultivating microorganisms (aerobic/anaerobic) in free-flowing liquid mediums like molasses and broths. This method is best adopted for microorganisms that demand high moisture, such as bacteria, and is most appropriate for the synthesis of extracellular enzymes. In this technique, selected microorganisms are cultured in a closed vessel containing a liquid fermentation medium. During fermentation, the desired products are secreted into the broth. In SmF, the substrates are quickly depleted; hence, they must be replenished and supplemented constantly. The three main modes of SmF processes are batch, continuous, and fed-batch. Over the past several years, SmF has gained immense importance in the production of commercial enzymes. For instance, 90% of commercial xylanases are produced using SmF; other enzymes, such as amylases and proteases, are also largely produced through SmF (Polizeli et al. 2005). The realm of SmF for enzyme production at the industrial scale has been facilitated by the strict control of process parameters, consistent productivity, ease of automation and downstream processing (Doriya et al. 2016).

6.2.2 Fermentation Parameters

The efficiency of fermentation-based processes is determined by a variety of parameters such as the selection of microorganism and substrate, temperature, pH, aeration, agitation, water activity and moisture contents, type of fermenter used, nature of solid substrate employed, etc. (Singhania et al. 2009; Farinas 2015). According to some authors, the composition of the fermentation medium and cultural conditions should be given more attention because the cost of the growth medium accounts for 30–40% of the production costs of cultural processes (Rangel et al. 2020). The most essential component affecting SSF operations is moisture, which comes from the nature of the solid substrate used. Moisture selection is influenced by the microorganisms used as well as the substrate's characteristics. Fungi require less moisture, 40–60% moisture may suffice, but substrate selection is influenced by a number of parameters, including cost and availability, and may necessitate the screening of numerous agro-industrial residues. Water interaction is one of the most important aspects of the SSF system that must be thoroughly examined. Microbial activity is influenced by the substrate's water activity (aw). (Webb 2017) The importance of aw has been intensively explored by researchers since it determines the sort of bacteria that can grow in SSF. The aw of the medium has been identified as a critical parameter for water and solute mass transport through microbial cells. Controlling this parameter could be utilized to alter a microorganism's metabolic production or excretion (Singhania et al. 2009; Costa et al. 2018). Inoculum preparation is a key step in a bioprocess since it can significantly influence production and yield. Therefore, inoculum stability and productivity should be monitored before transferring to the laboratory or pilot scale fermenter (Stanbury et al. 2013; Singh et al. 2017). It is necessary to maintain the organisms at their optimum temperature, pH, oxygen level, and substrate concentration which favour the maximum growth and product concentration in the fermentation broth.

6.3 Optimization Strategies for Enhanced Enzyme Production

Despite fermentation processes being used for many generations, the challenges associated with enzyme production such as low yield and inconsistent product quality are still a major drawback in the commercialization of any fermentation process (R. Singh et al. 2016). Various microorganisms have been reported to produce an array of enzymes, but enzyme yields obtained are insignificant and not up to the industrial mark. Moreover, scale-up is another critical issue associated with any fermentation process (Thiry and Cingolani 2002; Schmidt 2005). As fermentation technology has been applied for generations, its requirement for desired production capacity is high in the market that makes the fermentation process more demandable. For matching the market/industry's demand, a large quantity with high-quality production is needed. Optimizing process parameters, medium composition, and media conditions play a vital role in an industrial fermentation process as they are responsible for the generation, quantity, and quality of fermentation end-products as well as by-products consequently, affecting the total process economics. Therefore, in regards to increased benefits obtained from the fermentation process, it is crucial to focus on the optimization of the fermentation process that can be done by optimizing the process medium and conditions (Schmidt 2005). Over the past centuries, researchers have adopted medium optimization strategies, which is still one of the majorly investigated procedures to increase product yield. The pre-1970 era witnessed media optimization using classical approaches such as OFAT methods that are time-consuming, labour-intensive, and expansive (Desai et al. 2006; Singh et al. 2017). As discussed earlier, the fermentation process and cultural conditions play a pivotal role in industrial production, and optimizing the process parameters is very critical to achieving sustainable production with consistent product quality. With advancements in mathematical and statistical methods, medium and process optimization techniques have started gaining shape into new directions in terms of effectiveness, efficiency, economics, and robustness in giving the results. Figure 6.2 summarizes the various techniques and algorithms used for the optimization of medium composition and process parameters.

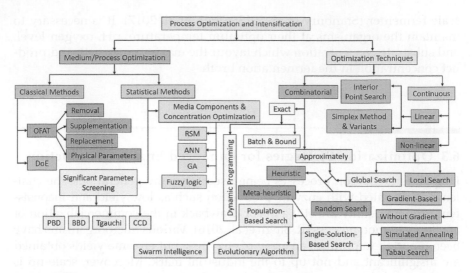

FIGURE 6.2
Various process optimization techniques and their classification.
Abbreviations: OFAT: One-factor at a time; DoE: Design of experiments; PBD: Plakett–Burman design; BBD: Box–Behnken design; CCD: Central composite design.

6.3.1 Medium Optimization Methods

Traditional/classical medium optimization techniques include OFAT and design of experiment (DoE) techniques for the optimization of medium components. In the OFAT approach, only one dependent variable or process parameter is changed over time while other parameters are kept constant. Based on the approach applied, the OFAT technique can be further classified into four categories:

1. Removal experiments where we remove one parameter from the set of parameters and study their effect on yield one-by-one

2. Supplementation experiments where we study the supplementation effects of nutrients on yields

3. Replacements experiments where the source of certain important substrates is varied to study its effect on product yield

4. Physical parameter experiments where the effect of biological and environmental parameters such as pH, temperature, agitation speed and aeration requirements are varied over time

This method is advantageous due to its simplicity in conducting a series of experiments and ease in results analysis. However, one cannot study the interactions of process parameters using the OFAT technique, which is a major drawback of this method. Alternative to classical techniques, one can use statistical techniques to optimize media components and their formulation process. Based on stages and objectives of optimization, statistical methods can be classified into two subclasses, viz. significant parameter screening and media component and concentration optimization. As the name suggests, significant parameter screening techniques are applied to identify those process parameters, which are very much crucial for the fermentation process. These include Plackett–Burman design (PBD), Box–Behnken design (BBD), Taguchi design, and central composite design (CCD). On the other hand, when one has identified significant parameters, they can further proceed to media components and concentration optimization techniques. This classification includes RSM, ANN, GA, and fuzzy logic. In this chapter, we will focus on the GA and its applications in fermentation process optimization. However, to read more about other algorithms and medium optimization techniques, readers are advised to go through previous literature, e.g., Singh et al. (2017) and Thiry and Cingolani (2002).

6.4 Genetic Algorithm as Optimization Technique for Fermentation Process

Optimization of the fermentation process contains multiple constraints such as expense, more labour work, open-ended and time-consuming processes that include multiple steps. Continuously, new strains and mutants are introduced in bioprocess industries and, therefore, there is the need to perform optimized experiments (Stanbury et al. 2013). As the fermentation process speeds, there will be a need for techniques to obtain high yield end-product generation at low cost which will compete with the conventional methods of fermentation processes. Currently, GAs which are non-statistical optimization techniques are used to optimize the fermentation process. The Genetic Algorithm is a stochastic optimization model that can be used to find an optimal or near-optimal solution by simulating natural evolution (Adeyemo and Enitan 2011). GAs are the compelling stochastic search algorithm associated with the theory of probability. The concept behind GA is genetics and the process of natural selection, where the historical information is used to take the algorithm to the best level. GAs are based on the "survival of the fittest" principle that means after repeated steps of mutation only the best individual will exist. For the identification of the ideal or best-fit solution set, the algorithm mimics the causes of evolution as in any natural population. At a single step, from the current population GAs randomly choose parents (some individual solutions) to reproduce the offspring for the next generation. As the repeat of the consecutive generation increases, the population evolves toward the most suitable solution. Likewise, GAs are responsible

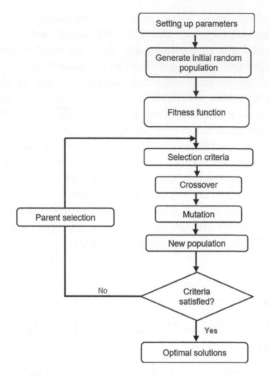

FIGURE 6.3
Workflow and steps involved in the application of genetic algorithm (GA) for parameter optimization in the fermentation process.

for the production of high yield products or solutions for optimization of fermentation using evolutionary biology, probability theory, and search algorithms. A detailed schematic workflow in GA is provided in Figure 6.3.

In GA, the initial solution set can be understood as an individual (with a specific chromosome) of a population. Two sets of chromosomes are selected as parents, and the offspring population is derived via the processes of recombination and mutation (Vasant and Barsoum 2009). Each chromosome consists of several genes that define a particular parameter, and each bit in a gene describes the value imparted to the parameter. The three processes that determine the next generation are "Selection", "Crossover", and "Mutation". Selection selects the parent chromosomes, and crossover combines them together to form the offspring as it happens in nature via recombination. Mutation introduces random changes in the chromosomes during each cycle (V. Singh et al. 2017). Selection usually occurs based on the fitness of the chromosomes defined by a fitness factor (Doganis et al. 2006). GA comes with several advantages over conventional optimization techniques.

It is capable of solving complex problems statements and can run them in parallel. GA can deal with various optimization methods, whose fitness (or objective) function can be of various types, viz. with random noise, linear/ nonlinear, stationary/non-stationary (that change over time), and continuous/discontinuous. Because multiple offspring in a population act like independent agents, the sample group (or population) can parallelly explore the search (or solution) space in multiple directions. Because of this feature, GA makes the process for finding optimal solution faster via parallelizing the algorithms. Furthermore, GAs are capable of manipulating multiple parameters and different groups of encoded strings simultaneously (Table 6.1).

TABLE 6.1

Microbial Sources of Various Enzymes and Their Optimization Using Conventional Optimization Techniques

Microbial Source	Enzyme	Applications	Optimization Method	Reference
Bacillus subtilis JK-79	L-glutaminase	Therapeutic enzyme	PBD and RSM	Kiruthika and Murugesan (2020)
Aeromonas veronii	L-glutaminase	Anti-cancer enzyme	PBD and CCD	Jesuraj et al. (2017)
Alcaligenes faecalis KLU102	L-glutaminase	Anti-cancer enzyme	RSM	Pandian et al. (2014)
Trichoderma reesei QM9414	Cellulases	Cellulolytic enzyme	CCRD	Sirohi et al. (2019)
Penicillium citrinum YL-1	Alkaline protease	Proteolytic	PBD and RSM	Xiao et al. (2015)
Streptomyces griseoloalbus	α-galactosidase	Glycoside hydrolase enzyme	RSM	Anisha et al. (2008)
Bacillus subtilis RSP-GLU MTCC 9727	L-glutaminase	Deamidation/aminohydrolase enzyme	RSM	Sathish et al. (2018)
Serratia marcescens NCIM 2919	L-asparaginase	Therapeutic enzyme	RSM	Ghosh et al. (2013)
Enterobacter aerogenes MTCC111	L-asparaginase	Therapeutic enzyme	RSM	Erva et al. (2017)
Bacillus sp. *VUVD101*	Lactase	Lignocellulolytic enzyme	PBD	Karlapudi et al. (2018)

Abbreviations: CCD – Central composite design, RSM – Response surface methodology, PBD – Plakett–Burman design

6.4.1 History of GA

GAs are considered to be global, parallel, and stochastic search algorithms. They are based on the principles of Darwin's evolutionary theory that deals with the idea of natural selection and genetics. This algorithm begins with a set of chromosomes referred to as a population. Offspring from a population are used to create the next population. The selection is done in such a way that the new population should be better than the previous one. These solutions are chosen according to their fitness means the more suitable they are the more probability for them to reproduce. This is continued until conditions are satisfied. Genetic algorithms are a booming area of AI as it comes under evolutionary computing. Evolutionary computing was initiated by Rechenberg in his work "Evolution strategies" in the 1960s. Other researchers further developed his idea. John Holland invented GAs and his students and colleagues developed them, which later led to the publication of Holland's book *Adaptation in Natural and Artificial Systems* in 1975. GA was used to evolve functions to execute particular tasks in 1992 by John Koza and referred to this method as "genetic programming" (GP). Since the discovery of the Simple Genetic Algorithm (SGA) by Holland (Holland 1992), several modifications of the same have been made in this. SGA usually solves problems by considering several different objectives as a linear set with a single objective. If some objectives mismatch, SGA results in more than one optimal solution. To overcome this issue associated with SGA, multi-objective GA (MOGA) was proposed (Dehuri and Mall 2006). Based on whether the Pareto principle is used or not, MOGA is classified into two groups. Non-Pareto-based models judge individual survival based on a calculated objective value such as Vector Evaluated GA (VEGA) and Hajela's and Lin's GA (HLGA). Pareto-based models check the dominance of individuals, where non-dominated chromosomes have a better survival capability. Examples of Pareto-based MOGAs include Niched Pareto GA (NPGA), Non-dominated Sorting GA (NSGA), and NSGA II (Horn et al. 1994; Zitzler and Thiele 1999). GA can also be combined with various Neural Network models, which guide the algorithm (Chen and Ramaswamy 2002).

6.4.2 Applications of GA to Optimize Fermentation Processes

The media composition is a major parameter that decides the products and their yields. Several efforts have been made to identify the optimum media for fermentation processes using GA. Marteijn et al. (2003) optimized the medium for the production of the *Helicoverpa zea* insect cells in a fed-batch culture for the increase of the cell density. A feed contained 11 different nutrients or parameters, and the concentration of each of them was described by a 5-digit code. Hence, the total length of the string (representing an individual) was 55-digits long. The calculations were done with a population size of 20, a crossover probability of 95%, and a mutation probability of 1% per bit. The optimized feed was tested in a bioreactor and led to a high cell density of 19.5 × 10⁶ cells/mL which is 550% higher than a commercial feed in a controlled bioreactor. Zheng et al. (2017) combined ANN, and GA optimized 16 nutrient concentrations in the fermentation process of wheat germ, which produces two anticancer compounds: 2, 6-dimethoxy-ρ-benzoquinone (DMBQ), and Methoxy-ρ-benzoquinone (MBQ). Each chromosome had 16 genes, and 30 such chromosomes were initially present. ANN-model-derived fitness function was used to select best-fit chromosomes for computing the next generation with a crossover probability of 0.9 and a mutation probability of 0.01. The optimized nutrient concentrations revealed a greater influence of micro- and macronutrients on MBQ and DMBQ production compared to the vitamins, with a total content of 0.939 mg/g. Zheng et al. (2016) also optimized other fermentation parameters such as fermentation time, temperature, agitation speed, and initial pH for MBQ, DBMQ, and flavonoid production from wheat germ. ANN was combined with NSGA II to obtain the results. Subba Rao et al. (2008) increased alkaline protease production by *Bacillus circulans*. Six factors were selected, including temperature, glucose concentration, pH, soya bean meal concentration, inoculum size, and medium volume. GA was combined with a feed-forward neural network (FFNN) to optimize 60 initial chromosomes of length 60, a mutation probability of 0.01, and a crossover probability of 0.9. The parameter values obtained enhanced enzyme yield by >2.5 times to 8320 Units. In any fermentation process, the temperature is another critical parameter. Maintaining the temperature within the optimum range is crucial to getting maximum output. de Andres-Toro et al. (1997) made a similar attempt in beer fermentation using GA in MATLAB. Each chromosome (or individual) contained 150 genes (for the 150 h study), with each gene having a value between 8°C and 18°C. One thousand two hundred such randomly generated individuals were included in the initial population. Each generation led to the addition of 400 new individuals with a crossover probability of 0.8 and a mutation probability of 0.008. The output provided a temperature profile for beer production and was even able to decrease fermentation time to 130 h. GA can be coupled with statistical modelling approaches such as ANN and RSM for effective optimization of input parameters in the fermentation process. For instance, amount of substrate, medium composition, extraction time, initial pH, fermentation time, agitation speed, balance of micronutrients, temperature, frequency, solid/liquid ratio (S/L), moisture content, and aeration rate can be optimized (Muthusamy et al. 2019). The use of GA gives more reliable and superior results over the conventional fermentation approaches because the results obtained from GA can be compared with the experimental results to confirm the optimization of parameters (Pedrozo et al. 2021). GA is used to investigate mathematical modelling to find a better model or for comparing the models, thereby finding optimal solutions for the problem. Parameters for the models are optimized using the GA. GAs have been employed to increase productivity during fermentation of various enzymes and bioactive compounds by comparing the kinetic models and their parameters as mentioned in Table 6.2. Mathematical kinetic Monod and Moser models were compared in a study for ethanol yield and recovery after fermentation using GA for CCBR (continuous conventional bioreactor) and CMBR (continuous membrane bioreactor) (Esfahanian et al. 2016). More applications of GA and hybrid GA models have been listed in Table 6.2.

TABLE 6.2

Microbial Sources of Various Enzymes and Their Optimization Using Various Hybrid Genetic Algorithm Techniques

Microbial Source	Enzyme	Applications	Optimization Method	Reference
Thermomyces lanuginosus VAPS-24	Xylanase	Lignocellulosic pretreatment, saccharification	CCD, GA-RSM	Kumar et al. (2017)
Bacillus circulans	Alkaline protease	Proteolytic	FFNN–GA	Subba Rao et al. (2008)
Bacillus licheniformis (MG5)	Alkaline protease	Proteolytic	GA-PSO	Ghovvati et al. (2015)
Streptomyces sp. *KP314280*	α-amylase	Starch liquefaction	(RSM-CCD) and (ANN-GA)	Ousaadi et al. (2021)
Streptomyces rimosus MTCC 10792	Cholesterol oxidase	Bioconversion of sterols, therapeutic enzyme	PBD and RSM-GA	Srivastava et al. (2018)
Penicillium Roqueforti ATCC 10110	Exoglucanase	Cellulolytic enzyme	Simplex-Centroid Design and ANN-GA	da Silva Nunes et al. (2020)
Trichoderma stromaticum AM7	Cellulases	Cellulolytic enzyme	ANN-GA	Bezerra et al. (2021)
Pleurotus ostreatus PVCRSP-7	Laccase	Delignification/ lignolytic enzyme	CCD, ANN-GA	Chiranjeevi et al. (2014)
Staphylococcus arlettae JPBW-1,	Lipases	lipolytic enzyme	CCD, ANN-GA	Chauhan et al. (2013)
Aspergillus terreus MTCC 1782	L-asparaginase	Deamidation/ amidohydrolase enzyme	RSM (PBD-CCD) and ANN-GA	Baskar and Renganathan (2012)

(Continued)

TABLE 6.2 (CONTINUED)

Microbial Source	Enzyme	Applications	Optimization Method	Reference
Rhizopus microsporus var. oligosporus CCT 3762	α-amylase	Starch liquefaction	ANN-GA	Fernández Núñez et al. (2017)
Penicillium roqueforti ATCC 10110	Lipase	Lipolytic enzyme	ANN-GA	Sales de Menezes et al. (2021)
Bacillus sp BGS	Alkaline protease	Proteolytic activity	RSM and GA	Moorthy and Baskar (2013)
Acinetobacter johnsonii	Novel dioxygenase (Diketone cleaving enzyme, Dke1)	Deoxygenation activity	GA	Hofer et al. (2004)
Bacillus subtilis RSP-GLU (MTCC 9727)	L- glutaminase	Deamidation/ amidohydrolase enzyme	ANN-GA	Sathish and Prakasham (2010)
Aspergillus niger	α-amylase	Starch saccharification	PSO and GA	Rajulapati and Lakshmi Narasu (2011)
Trichoderma reesei QM9414	Cellulase	Cellulolytic enzyme	GA	Sirohi et al. (2018)
Trichoderma harzianum	L-methioninase	Deamination and demethiolation of L-methionine	RSM and ANN-GA	Salim et al. (2019)
Aspergillus niger	α-amylase	Starch saccharification	RSM and GA	Rajulapati and Lakshmi Narasu (2011)
Pleurotus sajor-caju CCBIt 020	Manganese peroxidase (MnP) and laccase (Lac),	Lignin-modifying enzymes (LMEs)	RSM and ANN-GA	Vilar et al. (2021)
Pseudomonas sp. RAJR 044	Protease	Proteolytic activity	GA	Dutta et al. (2005)
Enterobacter aerogenes MTCC 111	L- asparaginase	Deamidation/ amidohydrolase enzyme	ANN-GA	Reddy and Babu (2017)
Aspergillus terreus MTCC 1782	L-asparaginase	Therapeutic enzyme	RSM and ANN-GA	Baskar and Renganathan (2012)
Pseudomonas fluorescens BL 915 ORF1 pSZexM2	Halogenase	Halogenation reaction	GA	Muffler et al. (2007)

(Continued)

TABLE 6.2 (CONTINUED)

Microbial Source	Enzyme	Applications	Optimization Method	Reference
Trichoderma reesei	Cellulases	Cellulolytic enzyme	RSM-GA	Saravanan et al. (2012)
Rhizopus oryzae NRRL 3562	Lipase	Lipolytic enzyme	GA and PSO	Garlapati et al. (2010)
Alternaria sp.	Antifungal exochitinase	Chitin degradation enzyme	DANN	Ismail et al. (2019)
Aspergillus sp.	Alkaline proteases	Proteolytic activity	RSM-GA	Mustefa Beyan et al. (2021)

Abbreviations: CCD – Central composite design, RSM – Response surface methodology, GA – Genetic algorithm, FFNN – feed-forward neural network, PSO – Particle swarm optimization, ANN – Artificial neural network, PBD – Plakett–Burman design

6.5 Problems and Bottlenecks in Optimization Techniques

Fermentation optimization involves numerous experiments which also include labour cost. It is unfortunate to see that the results obtained from the shake flask experiments (lab-scale studies) rarely match exactly with the fermenter studies. All lab-scale studies are majorly affected by four main weaknesses: uncontrollable pH, poor O_2 transfer capabilities, intermittent evaporation, and inefficient mixing during the fermentation process. Moreover, fermentation in an industrial-scale medium usually suffers from the problems of batch-to-batch variability, incessant fluctuations in biomass price, variations in the cost of product transportation, and problems associated with storage in bulk. Furthermore, microorganisms are very much dynamic in nature with a lot of internal control mechanisms, but most of the process optimization techniques treat them as blackbox or utilized solely for empirical data only. In this context, the next-generation process optimization techniques should take into account the metabolic pathway regulatory mechanism. Apart from this, the mutation rate of microbes should also be taken into consideration to account for the influence of medium components, as they affect the desired yield or product of interest. Future studies should also explore the effect of mutant strains (if available) in the medium optimization studies, as it might give us a different way to develop a new cost-effective process, where the use of a cheap medium can be accomplished. Most importantly, in contrast to liquid culture-based fermentation, a dearth of optimization studies has focussed on the solid or semi-SSF techniques which are the need of the hour owing to water scarcity. Furthermore, almost all the researchers while carrying out the optimization experiments, encounter the problem related to when one should stop optimizing a process or at which step optimization studies end. Yield is dependent on various physical and biochemical factors, therefore, designing an optimum medium for fermentation can be a never-ending problem. Many experts in this field always look out for new media components or media to achieve the desired yield.

Currently, GAs is applied to solve various search problems associated with optimization which involves probability function optimization, molecular docking, parameter fitting, scheduling, manufacturing, clustering, ML, etc. GA has been applied for simultaneous parameter optimization for microbial growth because microbial growth is a very important criterion in fermentation and it is regulated by several factors like competition, the release of toxic compounds, and metabolic activity (Pedrozo et al. 2021). However, the major disadvantages of GAs are that it lacks storage of the data created at a single step of the optimization process. The combination of GAs and ANN methods are more profitable in the optimization of the fermentation process. This combination of the method is based on the idea that after desired ANN and input time that is created in the scale of the independent framework can be improved using GAs. As compared to the GAs approach for optimization, the combination of GAs and ANN provide better fits to experimental information. This technique provides steady and systematic improvements of factors. At present, this combination of GA-ANN is becoming famous for the fermentation process and parameter optimization. Fermentation medium optimization is widely done using the ANN/GA-based approach for optimizing the macro- and micronutrients, and vitamins for specific product releases from the substrate (Singh et al. 2017; Zheng et al. 2017). Additionally, GA is applied for utilizing cheap Biological-Agro-industrial waste as substrates to produce high yields with reduced production costs (Sirohi et al. 2018). Genetic algorithms have been used in comparing various mathematical models including BP-ANN, SNN, RSM, PBD, and BBD for better optimization and higher production yields from the fermentation process. Despite many advantages over traditional or contemporary optimization algorithms, GAs also have some disadvantages. One needs to be extra careful while designing the objective function, the use of population size, the choice of the important parameters such as the rate of mutation and crossover, and the selection criteria of the new population should be carried out carefully. Any inappropriate choice will make it difficult for the algorithm to converge or it will simply produce meaningless results. Despite these drawbacks, genetic algorithms remain one of the most widely used optimization algorithms in modern nonlinear optimization.

6.6 Overview and Conclusions

For optimization of fermentation processes, GA is an evolutionary algorithm based on the process of natural selection of the involved parameters that finally provide the best solutions. GAs have been used for more than 40 years and are still being applied to gain optimized solutions where computational time does not matter much. They are the global optimization solutions, but focusing on the point of future scope, GAs can be effectively used, if it is combined with other latest computational algorithms that overcome the drawback of computational time, e.g., combinations like GAs-ANN and QIGA, and hybrid with memetic algorithms.

Bibliography

Adeyemo, Josiah, and Enitan, Abimbola. 2011. "Optimization of Fermentation Processes Using Evolutionary Algorithms – A Review." *Scientific Research and Essays* 6 (7): 1464–72. doi:10.5897/SRE11.106

Almadhoun, Wael, and Mohammad Hamdan. 2018. "Optimizing the Self-Organizing Team Size Using a Genetic Algorithm in Agile Practices." *Journal of Intelligent Systems* 29 (1): 1151–65. doi:10.1515/jisys-2018-0085

Al-Maqtari, Qais Ali, Waleed AL-Ansi, and Amer Ali Mahdi. 2019. "Microbial Enzymes Produced by Fermentation and Their Applications in the Food Industry – A Review." *International Journal of Agriculture Innovations and Research* 8 (1): 21.

Anisha, G. S., R. K. Sukumaran, and P. Prema. 2008. "Evaluation of Alpha-Galactosidase Biosynthesis by Streptomyces Griseoloalbus in Solid-State Fermentation Using Response Surface Methodology." *Letters in Applied Microbiology* 46 (3): 338–43. doi:10.1111/j.1472-765X.2008.02321.x

Balen, Manuela, Camila Silveira, Jadel M. Kratz, Cláudia M.O. Simões, Alexsandra Valério, Jorge L. Ninow, Leandro G. Nandi, Marco Di Luccio, and Débora de Oliveira. 2015. "Novozym® 435-Catalyzed Production of Ascorbyl Oleate in Organic Solvent Ultrasound-Assisted System." *Biocatalysis and Agricultural Biotechnology* 4 (4): 514–20. doi:10.1016/j.bcab.2015.08.008

Baskar, Gurunathan, and Sahadevan Renganathan. 2012. "Optimization of L-Asparaginase Production by Aspergillus Terreus MTCC 1782 Using Response Surface Methodology and Artificial Neural Network-Linked Genetic Algorithm." *Asia-Pacific Journal of Chemical Engineering* 7 (2): 212–20. doi:10.1002/apj.520

Bezerra, Camila Oliveira, Lucas Lima Carneiro, Elck Alemeida Carvalho, Thiago Pereira das Chagas, Lucas Ribeiro de Carvalho, Ana Paula Trovatti Uetanabaro, Gervásio Paulo da Silva, Erik Galvão Paranhos da Silva, and Andréa Miura da Costa. 2021. "Artificial Intelligence as a Combinatorial Optimization Strategy for Cellulase Production by Trichoderma Stromaticum AM7 Using Peach-Palm Waste Under Solid-State Fermentation." *BioEnergy Research* 14 (4): 1161–70. doi:10.1007/s12155-020-10234-4

Carboué, Quentin, Marie-Stéphane Tranier, Isabelle Perraud-Gaime, and Sevastianos Roussos. 2017. "Production of Microbial Enzymes by Solid-State Fermentation for Food Applications." In *Microbial Enzyme Technology in Food Applications*, edited by Ramesh C. Ray and Cristina M. Rosell, 1st ed., 437–51. Boca Raton, FL:

CRC Press, [2016] | Series: Food biology series | "A science publishers book.": CRC Press. doi:10.1201/9781315368405-28

Chauhan, Mamta, Rajinder Singh Chauhan, and Vijay Kumar Garlapati. 2013. "Modelling and Optimization Studies on a Novel Lipase Production by *Staphylococcus Arlettae* through Submerged Fermentation." *Enzyme Research* 2013 (December): 1–8. doi:10.1155/2013/353954

Chen, C.R., and H.S. Ramaswamy. 2002. "Modeling and Optimization of Variable Retort Temperature (VRT) Thermal Processing Using Coupled Neural Networks and Genetic Algorithms." *Journal of Food Engineering* 53 (3): 209–20. doi:10.1016/S0260-8774(01)00159-5

Chiranjeevi, Potu Venkata, Moses Rajasekara Pandian, and Sathish Thadikamala. 2014. "Integration of Artificial Neural Network Modeling and Genetic Algorithm Approach for Enrichment of Laccase Production in Solid State Fermentation by Pleurotus Ostreatus." *BioResources* 9 (2): 2459–70.

Costa, Jorge A.V., Helen Treichel, Vinod Kumar, and Ashok Pandey. 2018. "Advances in Solid-State Fermentation." In: A. Pandey. C. Larroche, and C. R. Socco (eds), *Current Developments in Biotechnology and Bioengineering*, 1–17. Elsevier. doi:10.1016/B978-0-444-63990-5.00001-3

da Silva Nunes, Nájila, Lucas Lima Carneiro, Luiz Henrique Sales de Menezes, Marise Silva de Carvalho, Adriana Bispo Pimentel, Tatielle Pereira Silva, Clissiane Soares Viana Pacheco, et al. 2020. "Simplex-Centroid Design and Artificial Neural Network-Genetic Algorithm for the Optimization of Exoglucanase Production by Penicillium Roqueforti ATCC 10110 Through Solid-State Fermentation Using a Blend of Agroindustrial Wastes." *BioEnergy Research* 13 (4): 1130–43. doi:10.1007/s12155-020-10157-0

de Andres-Toro, B., J. M. Giron-Sierra, J. A. Lopez-Orozco, and C. Fernandez-Conde. 1997. "Evolutionary Optimization of an Industrial Batch Fermentation Process." In *1997 European Control Conference (ECC)*, 2401–6. Brussels: IEEE. doi:10.23919/ECC.1997.7082465

Dehuri, S., and R. Mall. 2006. "Predictive and Comprehensible Rule Discovery Using a Multi-Objective Genetic Algorithm." *Knowledge-Based Systems* 19 (6): 413–21. doi:10.1016/j.knosys.2006.03.004

Desai, K.M., S.K. Akolkar, Y.P. Badhe, S.S. Tambe, and S.S. Lele. 2006. "Optimization of Fermentation Media for Exopolysaccharide Production from Lactobacillus Plantarum Using Artificial Intelligence-Based Techniques." *Process Biochemistry* 41 (8): 1842–48. doi:10.1016/j.procbio.2006.03.037

Doganis, Philip, Alex Alexandridis, Panagiotis Patrinos, and Haralambos Sarimveis. 2006. "Time Series Sales Forecasting for Short Shelf-Life Food Products Based on Artificial Neural Networks and Evolutionary Computing." *Journal of Food Engineering* 75 (2): 196–204. doi:10.1016/j.jfoodeng.2005.03.056

Doriya, K., N. Jose, M. Gowda, and D.S. Kumar. 2016. "Solid-State Fermentation vs Submerged Fermentation for the Production of l-Asparaginase." *Advances in Food and Nutrition Research* 78: 115–35. doi:10.1016/bs.afnr.2016.05.003

Dutta, Jayati Ray, Pranab Kumar Dutta, and Rintu Banerjee. 2005. "Modeling and Optimization of Protease Production by a Newly Isolated Pseudomonas Sp. Using a Genetic Algorithm." *Process Biochemistry* 2 (40): 879–84. doi:10.1016/j.procbio.2004.02.013

Erva, Rajeswara Reddy, Ajgebi Nath Goswami, Priyanka Suman, Ravali Vedanabhatla, and Satish Babu Rajulapati. 2017. "Optimization of L-Asparaginase Production from Novel Enterobacter Sp., by Submerged Fermentation Using Response Surface Methodology." *Preparative Biochemistry & Biotechnology* 47 (3): 219–28. doi:10.1080/10826068.2016.1201683

Esfahanian, Mehri, Ali Shokuhi Rad, Saeed Khoshhal, Ghasem Najafpour, and Behnam Asghari. 2016. "Mathematical Modeling of Continuous Ethanol Fermentation in a Membrane Bioreactor by Pervaporation Compared to Conventional System: Genetic Algorithm." *Bioresource Technology* 212 (July): 62–71. doi:10.1016/j.biortech.2016.04.022

Farinas, Cristiane S. 2015. "Developments in Solid-State Fermentation for the Production of Biomass-Degrading Enzymes for the Bioenergy Sector." *Renewable and Sustainable Energy Reviews* 52 (December): 179–88. doi:10.1016/j.rser.2015.07.092

Garlapati, Vijay Kumar, Pandu Ranga Vundavilli, and Rintu Banerjee. 2010. "Evaluation of Lipase Production by Genetic Algorithm and Particle Swarm Optimization and Their Comparative Study." *Applied Biochemistry and Biotechnology* 162 (5): 1350–61. doi:10.1007/s12010-009-8895-2

Ghosh, Sangita, Sonam Murthy, Sharmila Govindasamy, and Muthukumaran Chandrasekaran. 2013. "Optimization of L-Asparaginase Production by Serratia Marcescens (NCIM 2919) under Solid State Fermentation Using Coconut Oil Cake." *Sustainable Chemical Processes* 1 (1): 9. doi:10.1186/2043-7129-1-9

Ghovvati, Mahsa, Gholam Khayati, Hossein Attar, and Ali Vaziri. 2015. "Comparison across Growth Kinetic Models of Alkaline Protease Production in Batch and Fed-Batch Fermentation Using Hybrid Genetic Algorithm and Particle Swarm Optimization." *Biotechnology & Biotechnological Equipment* 29 (6): 1216–25. doi:10.1080/13102818.2015.1077686

Gustavo, Fernández Núñez Eutimio, Augusto Cesar Barchi, Shuri Ito, Bruna Escaramboni, Rondinelli Donizetti Herculano, Cassia Roberta Malacrida Mayer, and Pedro de Oliva Neto. 2017. "Artificial Intelligence Approach for High Level Production of Amylase Using Rhizopus Microsporus Var. Oligosporus and Different Agro-Industrial Wastes." *Journal of Chemical Technology & Biotechnology* 92 (3): 684–92. doi:10.1002/jctb.5054

Hofer, Hannes, Thomas Mandl, and Walter Steiner. 2004. "Diketone Cleaving Enzyme Dke1 Production by Acinetobacter Johnsonii--Optimization of Fermentation Conditions." *Journal of Biotechnology* 107 (1): 73–81. doi:10.1016/j.jbiotec.2003.09.010

Holland, John H. 1992. "Genetic Algorithms." *Scientific American* 267 (1): 66–73.

Horn, J., N. Nafpliotis, and D.E. Goldberg. 1994. "A Niched Pareto Genetic Algorithm for Multiobjective Optimization." In *Proceedings of the First IEEE Conference on Evolutionary Computation. IEEE World Congress on Computational Intelligence*, 82–87. Orlando, FL, USA: IEEE. doi:10.1109/ICEC.1994.350037

Industrial Enzymes Market Global Outlook, Trends, and Forecast to 2026 | MarketsandMarkets. 2022. Accessed February 20. https://www.marketsandmarkets.com/Market-Reports/industrial-enzymes-market-237327836.html.

Ismail, Shaymaa A, Ahmed Serwa, Amira Abood, Bahgat Fayed, Siham A Ismail, and Amal M Hashem. 2019. "A Study of the Use of Deep Artificial Neural Network in the Optimization of the Production of Antifungal Exochitinase Compared with the Response Surface Methodology." *Jordan Journal of Biological Sciences* 12 (5): 9.

Jesuraj, S. Aravinth Vijay, Md Moklesur Rahman Sarker, Long Chiau Ming, S. Marylin Jeya Praya, M. Ravikumar, and Wong Tin Wui. 2017. "Enhancement of the Production of L-Glutaminase, an Anticancer Enzyme, from Aeromonas Veronii by Adaptive and Induced Mutation Techniques." *PLOS ONE* 12 (8): e0181745. doi:10.1371/journal.pone.0181745

Karlapudi, Abraham Peele, S. Krupanidhi, Reddy E. Rajeswara, M. Indira, N. Bobby Md, and T.C. Venkateswarulu. 2018. "Plackett-Burman Design for Screening of Process Components and Their Effects on Production of Lactase by Newly Isolated Bacillus Sp. VUVD101 Strain from Dairy Effluent." *Beni-Suef University Journal of Basic and Applied Sciences* 7 (4): 543–46. doi:10.1016/j.bjbas.2018.06.006

Kiruthika, Jambulingam, and Saranya Murugesan. 2020. "Studies on Optimization of L-Glutaminase Production under Submerged Fermentation from Marine Bacillus Subtilis JK-79." *African Journal of Microbiology Research* 14 (1). Academic Journals: 16–24. doi:10.5897/AJMR2019.9107

Kumar, Karan, Kuldeep Roy, and Vijayanand S. Moholkar. 2021. "Mechanistic Investigations in Sonoenzymatic Synthesis of N-Butyl Levulinate." *Process Biochemistry* 111 (December): 147–58. doi:10.1016/j.procbio.2021.09.005

Kumar, Vishal, Deepak Chhabra, and Pratyoosh Shukla. 2017. "Xylanase Production from Thermomyces Lanuginosus VAPS-24 Using Low Cost Agro-Industrial Residues via Hybrid Optimization Tools and Its Potential Use for Saccharification." *Bioresource Technology* 243 (November): 1009–19. doi:10.1016/j.biortech.2017.07.094

Lee, Se Yeon, and Chae Hun Ra. 2021. "Comparison of Liquid and Solid-State Fermentation Processes for the Production of Enzymes and Beta-Glucan from Hulled Barley." *Journal of Microbiology and Biotechnology*, December. doi:10.4014/jmb.2111.11002

Malani, Ritesh S., Sachin B. Umriwad, Karan Kumar, Arun Goyal, and Vijayanand S. Moholkar. 2019. "Ultrasound–Assisted Enzymatic Biodiesel Production Using Blended Feedstock of Non–Edible Oils: Kinetic Analysis." *Energy Conversion and Management* 188 (May): 142–50. doi:10.1016/j.enconman.2019.03.052

Marteijn, R. C. L., O. Jurrius, J. Dhont, C. D. de Gooijer, J. Tramper, and D. E. Martens. 2003. "Optimization of a Feed Medium for Fed-Batch Culture of Insect Cells Using a Genetic Algorithm." *Biotechnology and Bioengineering* 81 (3): 269–78. doi:10.1002/bit.10465

Mondal, Payel, Anup Kumar Sadhukhan, Amit Ganguly, and Parthapratim Gupta. 2021. "Optimization of Process Parameters for Bio-Enzymatic and Enzymatic Saccharification of Waste Broken Rice for Ethanol Production Using Response Surface Methodology and Artificial Neural Network–Genetic Algorithm." *3 Biotech* 11 (1): 28. doi:10.1007/s13205-020-02553-2

Moorthy, Innasi Muthu Ganesh, and Rajoo Baskar. 2013. "Statistical Modeling and Optimization of Alkaline Protease Production from a Newly Isolated Alkalophilic Bacillus Species BGS Using Response Surface Methodology and Genetic Algorithm." *Preparative Biochemistry & Biotechnology* 43 (3): 293–314. doi:10.1080/10826068.2012.719850

Muffler, Kai, Marco Retzlaff, Karl-Heinz van Pée, and Roland Ulber. 2007. "Optimisation of Halogenase Enzyme Activity by Application of a Genetic Algorithm." *Journal of Biotechnology* 127 (3): 425–33. doi:10.1016/j.jbiotec.2006.07.008

Mustefa Beyan, Surafel, S. Venkatesa Prabhu, Tsegazeab K. Mumecha, and Mesfin T. Gemeda. 2021. "Production of Alkaline Proteases Using Aspergillus Sp. Isolated from Injera: RSM-GA Based Process Optimization and Enzyme Kinetics Aspect." *Current Microbiology* 78 (5): 1823–34. doi:10.1007/s00284-021-02446-4

Muthusamy, Shanmugaprakash, Lakshmi Priya Manickam, Venkateshprabhu Murugesan, Chandrasekaran Muthukumaran, and Arivalagan Pugazhendhi. 2019. "Pectin Extraction from Helianthus Annuus (Sunflower) Heads Using RSM and ANN Modelling by a Genetic Algorithm Approach." *International Journal of Biological Macromolecules* 124 (March): 750–58. doi:10.1016/j.ijbiomac.2018.11.036

Ousaadi, Mouna Imene, Fateh Merouane, Mohammed Berkani, Fares Almomani, Yasser Vasseghian, and Mahmoud Kitouni. 2021. "Valorization and Optimization of Agro-Industrial Orange Waste for the Production of Enzyme by Halophilic Streptomyces Sp." *Environmental Research* 201 (October): 111494. doi:10.1016/j.envres.2021.111494

Pandian, Sureshbabu Ram Kumar, V. Deepak, S. Sivasubramaniam, H. Nellaiah, and K. Sundar. 2014. "Optimization and Purification of Anticancer Enzyme L-Glutaminase from Alcaligenes Faecalis KLU102." *Biologia*. doi:10.2478/s11756-014-0486-1

Pedrozo, Hector A., Andrea M. Dallagnol, and Carlos E. Schvezov. 2021. "Genetic Algorithm Applied to Simultaneous Parameter Estimation in Bacterial Growth." *Journal of Bioinformatics and Computational Biology* 19 (1): 2050045. doi:10.1142/S0219720020500456

Polizeli, M. L. T. M., A. C. S. Rizzatti, R. Monti, H. F. Terenzi, J. A. Jorge, and D. S. Amorim. 2005. "Xylanases from fungi: properties and industrial applications." *Applied Microbiology and Biotechnology* 67 (5). doi:10.1007/s00253-005-1904-7

Rajulapati, Satish Babu, and M. Lakshmi Narasu. 2011. "Studies on A-Amylase and Ethanol Production from Spoiled Starch Rich Vegetables and Multi Objective Optimization by P.S.O and Genetic Algorithm." In *International Conference on Green Technology and Environmental Conservation (GTEC-2011)*, 37–41. doi:10.1109/GTEC.2011.6167638

Rangel, Albert E. T., Jorge Mario Gómez Ramírez, and Andrés Fernando González Barrios. 2020. "From Industrial By-products to Value-added Compounds: The Design of Efficient Microbial Cell Factories by Coupling Systems Metabolic Engineering and Bioprocesses." *Biofuels, Bioproducts and Biorefining* 14 (6): 1228–38. doi:10.1002/bbb.2127

Reddy, Erva Rajeswara, and Rajulapati Satish Babu. 2017. "Neural Network Modeling and Genetic Algorithm Optimization Strategy for the Production of L-Asparaginase from Novel Enterobacter Sp." *Journal of Pharmaceutical Sciences* 9: 7.

Roeva, O. 2006. "A Modified Genetic Algorithm for a Parameter Identification of Fermentation Processes." *Biotechnology & Biotechnological Equipment* 20 (1): 202–9. doi:10.1080/13102818.2006.10817333

Sadh, Pardeep Kumar, Surekha Duhan, and Joginder Singh Duhan. 2018. "Agro-Industrial Wastes and Their Utilization Using Solid State Fermentation: A Review." *Bioresources and Bioprocessing* 5 (1): 1. doi:10.1186/s40643-017-0187-z

Sales de Menezes, Luiz Henrique, Lucas Lima Carneiro, Iasnaia Maria de Carvalho Tavares, Pedro Henrique Santos, Thiago Pereira das Chagas, Adriano Aguiar Mendes, Erik Galvão Paranhos da Silva, Marcelo Franco, and Julieta Rangel de Oliveira. 2021. "Artificial Neural Network Hybridized with a Genetic Algorithm for Optimization of Lipase Production from Penicillium Roqueforti ATCC 10110 in Solid-State Fermentation." *Biocatalysis and Agricultural Biotechnology* 31 (January): 101885. doi:10.1016/j.bcab.2020.101885

Salim, Nisha, A. Santhiagu, and K. Joji. 2019. "Process Modeling and Optimization of High Yielding L-Methioninase from a Newly Isolated Trichoderma Harzianum Using Response Surface Methodology and Artificial Neural Network Coupled Genetic Algorithm." *Biocatalysis and Agricultural Biotechnology*. Elsevier Ltd. doi:10.1016/j.bcab.2018.11.032

Saravanan, P., R. Muthuvelayudham, R. Rajesh Kannan, and T. Viruthagiri. 2012. "Optimization of Cellulase Production Using Trichoderma Reesei by RSM and Comparison with Genetic Algorithm." *Frontiers of Chemical Science and Engineering* 6 (4): 443–52. doi:10.1007/s11705-012-1225-1

Sathish, Thadikamala, Devarapalli Kezia, P. V. Bramhachari, and Reddy Shetty Prakasham. 2018. "Multi-Objective Based Superimposed Optimization Method for Enhancement of 1 -Glutaminase Production by Bacillus Subtilis RSP-GLU." *Karbala International Journal of Modern Science* 4 (1): 50–60. doi:10.1016/j.kijoms.2017.10.006

Sathish, Thadikamala, and Reddy Shetty Prakasham. 2010. "Enrichment of Glutaminase Production by Bacillus Subtilis RSP-GLU in Submerged Cultivation Based on Neural Network—Genetic Algorithm Approach." *Journal of Chemical Technology & Biotechnology* 85 (1): 50–58. doi:10.1002/jctb.2267

Schmidt, F. R. 2005. "Optimization and Scale up of Industrial Fermentation Processes." *Applied Microbiology and Biotechnology* 68 (4): 425–35. doi:10.1007/s00253-005-0003-0

Singh, Neha, Karan Kumar, Arun Goyal, and Vijayanand S. Moholkar. 2022. "Ultrasound-Assisted Biodiesel Synthesis by in–Situ Transesterification of Microalgal Biomass: Optimization and Kinetic Analysis." *Algal Research* 61 (January): 102582. doi:10.1016/j.algal.2021.102582

Singh, Rajendra, Manoj Kumar, Anshumali Mittal, and Praveen Kumar Mehta. 2016. "Microbial Enzymes: Industrial Progress in 21st Century." *3 Biotech* 6 (2): 174. doi:10.1007/s13205-016-0485-8

Singh, Vineeta, Shafiul Haque, Ram Niwas, Akansha Srivastava, Mukesh Pasupuleti, and C. K. M. Tripathi. 2017. "Strategies for Fermentation Medium Optimization: An In-Depth Review." *Frontiers in Microbiology* 7 (January). doi:10.3389/fmicb.2016.02087

Singhania, Reeta Rani, Anil Kumar Patel, Carlos R. Soccol, and Ashok Pandey. 2009. "Recent Advances in Solid-State Fermentation." *Biochemical Engineering Journal* 44 (1): 13–18. doi:10.1016/j.bej.2008.10.019

Sirohi, Ranjna, Anupama Singh, Ayon Tarafdar, and N.C. Shahi. 2018. "Application of Genetic Algorithm in Modelling and Optimization of Cellulase Production." *Bioresource Technology* 270 (December): 751–54. doi:10.1016/j.biortech.2018.09.105

Sirohi, Ranjna, Anupama Singh, Ayon Tarafdar, Navin Chandra Shahi, Ashok Kumar Verma, and Anurag Kushwaha. 2019. "Cellulase Production from Pre-Treated Pea Hulls Using Trichoderma Reesei Under Submerged Fermentation." *Waste and Biomass Valorization*. Springer, Netherlands. doi:10.1007/s12649-018-0271-4

Soccol, Carlos Ricardo, Eduardo Scopel Ferreira da Costa, Luiz Alberto Junior Letti, Susan Grace Karp, Adenise Lorenci Woiciechowski, and Luciana Porto de Souza Vandenberghe. 2017. "Recent Developments and Innovations in Solid State Fermentation." *Biotechnology Research and Innovation* 1 (1): 52–71. doi:10.1016/j.biori.2017.01.002

Srivastava, Akanksha, Vineeta Singh, Shafiul Haque, Smriti Pandey, Manisha Mishra, Arshad Jawed, P. K. Shukla, P. K. Singh, and C. K. M. Tripathi. 2018. "Response Surface Methodology-Genetic Algorithm Based Medium Optimization, Purification, and Characterization of Cholesterol Oxidase from Streptomyces Rimosus." *Scientific Reports* 8 (1): 10913. doi:10.1038/s41598-018-29241-9

Stanbury, Peter F., Allan Whitaker, and Stephen J. Hall. 2013. *Principles of Fermentation Technology*. Elsevier.

Subba Rao, Ch., T. Sathish, M. Mahalaxmi, G. Suvarna Laxmi, R. Sreenivas Rao, and R.S. Prakasham. 2008. "Modelling and Optimization of Fermentation Factors for Enhancement of Alkaline Protease Production by Isolated Bacillus Circulans Using Feed-Forward Neural Network and Genetic Algorithm." *Journal of Applied Microbiology* 104 (3): 889–98. doi:10.1111/j.1365-2672.2007.03605.x

Thiry, Michel, and Doriano Cingolani. 2002. "Optimizing Scale-up Fermentation Processes." *Trends in Biotechnology* 20 (3): 103–5. doi:10.1016/S0167-7799(02)01913-3

Vasant, Pandian, and Nader Barsoum. 2009. "Hybrid Genetic Algorithms and Line Search Method for Industrial Production Planning with Non-Linear Fitness Function." *Engineering Applications of Artificial Intelligence* 22 (4–5): 767–77. doi:10.1016/j.engappai.2009.03.010

Vilar, Débora S., Clara D. Fernandes, Victor R. S. Nascimento, Nádia H. Torres, Manuela S. Leite, Ram Naresh Bharagava, Muhammad Bilal, Giancarlo R. Salazar-Banda, Katlin I. Barrios Eguiluz, and Luiz Fernando Romanholo Ferreira. 2021. "Hyper-Production Optimization of Fungal Oxidative Green Enzymes Using Citrus Low-Cost Byproduct." *Journal of Environmental Chemical Engineering* 9 (1): 105013. doi:10.1016/j.jece.2020.105013

Webb, Colin. 2017. "Design Aspects of Solid State Fermentation as Applied to Microbial Bioprocessing." *Journal of Applied Biotechnology & Bioengineering* 4 (1). doi:10.15406/jabb.2017.04.00094

Xiao, Yun-Zhu, Duan-Kai Wu, Si-Yang Zhao, Wei-Min Lin, and Xiang-Yang Gao. 2015. "Statistical Optimization of Alkaline Protease Production from Penicillium Citrinum YL-1 Under Solid-State Fermentation." *Preparative Biochemistry & Biotechnology* 45 (5): 447–62. doi:10.1080/10826068.2014.923450

Zheng, Ziyi, Xiaona Guo, Kexue Zhu, Wei Peng, and Huiming Zhou. 2016. "The Optimization of the Fermentation Process of Wheat Germ for Flavonoids and Two Benzoquinones Using EKF-ANN and NSGA-II." *RSC Advances* 6 (59): 53821–29. doi:10.1039/C5RA27004A

Zheng, Zi-Yi, Xiao-Na Guo, Ke-Xue Zhu, Wei Peng, and Hui-Ming Zhou. 2017. "Artificial Neural Network – Genetic Algorithm to Optimize Wheat Germ Fermentation Condition: Application to the Production of Two Anti-Tumor Benzoquinones." *Food Chemistry* 227 (July): 264–70. doi:10.1016/j.foodchem.2017.01.077

Zitzler, E., and L. Thiele. 1999. "Multiobjective Evolutionary Algorithms: A Comparative Case Study and the Strength Pareto Approach." *IEEE Transactions on Evolutionary Computation* 3 (4): 257–71. doi:10.1109/4235.797969

7

Optimization of Process Parameters of Various Classes of Enzymes Using Artificial Neural Network

Rajeev Kumar

Dayananda Sagar University, Bangalore, India

S. M. Veena and C. Sowmya

Sapthagiri College of Engineering, Bangalore, India

Ajay Nair and Archana S. Rao

Dayananda Sagar University, Bangalore, India

Uday Muddapur

KLE Tach University, Hubli, India

K. S. Anantharaju

Dayananda Sagar College of Engineering, Bangalore, India

Sunil S. More

Dayananda Sagar University, Bangalore, India

CONTENTS

7.1 Introduction

Enzymes are biological catalysts that facilitate chemical reactions under feasible physio-chemical conditions. The enzymes or their products are used in many applications for human use. The industrial production of enzymes witnesses constant growth. Microorganisms are the preferred sources of industrial enzymes. Their uses are increasing in many industries such as dairy, baking, beverage, animal feed, pulp and paper, detergent, leather cosmetics, organic synthesis, and waste management (Singh et al. 2016). The enzyme industry employs different classes of enzymes such as amylases, proteases, and lipases. The production cost of the enzyme is the most important in the growth and sustainability of the enzyme industry (Klein-Marcuschamer et al. 2012). There are several factors that determine the production cost of enzymes. The most important factors (process parameters) determining the production of enzymes are media compositions and fermentation operating conditions (Kennedy and Krouse 1999). Therefore, optimization of the significant process parameters is a desirable and necessary strategy for enzyme production.

Many strategies are being used to optimize the process parameters to produce enzymes (Singh et al. 2017). The simplest method for optimization is a one-at-a-time method. However, this method is time-consuming and is not feasible for the complex data sets used in the optimization process. With the development of response surface methodology (RSM) in 1951, statistical methods became popular and are widely used in the optimization process. RSM in association with design of experiments (DOE) has been successfully implemented for optimizing bioprocesses (Mandenius and Brundin 2008).

Machine learning (ML) models are sub-domain of artificial intelligence (AI). The key motivation for artificial intelligence is to make the computer an intelligent machine that can help in the decision-making process. Artificial neural networks (ANN) have been developed in the pursuit of developing AI. Human brain functioning is the guiding principle for the development of ANN. ANN can be used to understand the hidden pattern in a complex data set. The enzyme production process generates complex data set involving complex relationships among different process parameters. There has been a growing interest in using ANN for optimizing the process parameters for enzyme production. The current interest in using ANN for the optimization of process parameters is the result of the development of optimization algorithms techniques and the increased power of computing power of computers. The image processing technology has been successfully integrated with ANN techniques. This integration has been used in many AI-driven technologies including the driverless car. These successes can be replicated in the operation of the fermentation industry for increasing enzyme production.

This chapter aims to explore the application of the ANN model for the optimization of process parameters for the optimization of the production of various classes of the enzyme. In this chapter, the basic strategies of optimization have been discussed. A detailed description of the ANN method is given with an emphasis on the application of ANN in the context of optimization of the process parameters for enzyme production. A brief description of RSM is given and this methodology is compared with ANN. The focus here is to address the question of whether ANN can provide an alternative method for the optimization of process parameters. Finally, the application of ANN in the optimization of process parameters for enzyme production is discussed in the conclusion.

7.2 Strategy to Solve Optimization Problems

Optimization is the process of obtaining the best result by choosing the combinations of inputs under the given circumstance. The process of optimization is relevant where the outcome of the process depends on several factors. Using optimization, the outcome is manipulated by changing the factors which control the outcome. A process is optimized with the aim of finding the desired outcome which often involves either maximizing or minimizing the outcome. These are called the goals of optimization and the factors which are modulated are called the decision variables. The biological systems present typical examples of optimized systems. For example, the infoldings of the small intestine of animals, including humans, are evolved to maximize the surface area for absorbing the nutrients from the gut. Similarly, the evolution of an organism illustrates the examples of optimization, and these serve as a motivation to develop more efficient systems for human use. Optimization problems can be constrained or unconstrained. In constrained optimization, optimization is done subject to some conditions. In unconstrained optimization, there are no conditions.

The yield of the enzyme depends on several factors including media composition, operating condition, and the functioning of reactors. Therefore, the optimization of the enzyme process is necessary for enzyme production.

Modeling and simulations are important ways to address the optimization of the process parameters for enzyme production. Kinetic modeling is a mechanistic approach to optimizing the process parameters. However, the identification of the parameters is difficult to determine and is not practical in all cases. Therefore, we need an alternative way to get the optimum combination of process parameters.

Mathematical models are necessary for the optimization process. These models can be simple or complex. A simple empirical mathematical function can serve as a simple model and can be used for optimization. The complex model can be derived from the data or developed based on system characteristics. There are three broad strategies to obtain solutions for optimization problems. The difference among these strategies lies in the characteristics of variables, optimization function, and constraints (Balaman 2019).

A. Linear programming (LP): Linear programming requires the optimization function and the constraints to be linear. There should be a proportionality relationship between the variables and the objective function. The aim of this method is to get a set of optimal values for the decision variables by maximizing or minimizing a linear objective function. Depending upon the number of constraints, linear programming can be small (less than 1000 constraints), medium (between 1000 and 2000 constraints), and large (more than 2000 constraints) (Dantzig and Thapa 1997).

   The main requirement in the application of linear programming is that systems must operate in a linear range. Möller et al. (2018) used linear programming to understand the behavior of cancer cells in the utilization of glucose. Cancer cells convert glucose into lactate even in the presence of sufficient oxygen. This is known as the 'Warburg effect'. Möller et al. (2018) developed a minimal model and used linear programming with maximizing the ATP production rate as an optimization function and with different combinations of constraints on enzymes. Using this approach, the model predicted the oxidation of glucose under the reduction in glucose concentration or uptake process. Torres et al. (1996) identified the requirements for the substrate pool and the range of enzyme activity and several enzymes to increase the glycolytic flux for the production of citric acid in *Aspergillus niger* by formulating a linear optimization problem. They used linear programming and identified seven or more enzymes that could increase citric acid production by 3-fold.

B. Dynamic programming (DP): Time is an important factor in this strategy of obtaining a solution for an optimization problem. This strategy can be applied in situations such as maximizing the return of an investment. In order to maximize the return of an investment, money has to be allocated in different portfolios. A decision needs to be taken when some return is obtained after some time which again needs to be reallocated to increase the final return. Dynamic programming is characterized by two features – recursive operation of the execution of the program and multi-stage decision-making process (Gluss 1961). In this method, the solution is improved after each stage of decision-making. Ilkova and Petrov (2008) applied dynamic programming in the estimation of optimal initial concentration of biomass, substrate, and feed substrate in a fed-batch fermentation process using laboratory *Escherichia coli*. They also reported the reduction in the iteration of obtaining the desired optimization function.

C. Non-linear programming (NLP): Non-linear programming addresses the optimization problems where the optimization function and constrain operate in a non-linear fashion. Such situations are common in industrial operations. One of the common issues is finding the global minima during an optimization process. NLP is primarily developed to address the problem of finding the global minima. The common strategy adopted in NLP is searching the global minima by iterative approach. The NLP method form a good strategy to solve the complex optimization problem in combination with linear programming methods (Luenberger 1973). Evolutionary programming and simulated annealing are common examples of NLP.

Machine learning methods are being increasingly used in solving optimizing problems (Sun et al. 2019). The solution to an optimization problem is built in the formulation of a machine learning model. The main learning processes – classification and clustering – are the optimization techniques. The linear regression model and decision tree algorithms are employed in solving the optimization problem of a relatively simple nature. The more complex optimization problems can be addressed by ANNs.

7.3 Description and Architecture of Artificial Neural Networks

Warren McCulloch and Water Pitts opened the field of ANNs in 1943 (McCulloch and Pitts 1943). They proposed a simple model of a biological neuron. Their model was formulated on Boolean logic and the model output was in a binary form. This model was less flexible as the model used to input in binary form and could perform simple tasks. Frank Rosenblatt put forward a refined version of the neuron model and called it a perceptron (Rosenblatt 1962). Perceptron was a single-layer neural model which could take real input. This model could learn using different algorithms. By the 1960s, the ANN was used to understand the simple text and human faces. The model was able to differentiate between male and female faces. Minski and Papert (1969) identified the limitation of perceptron in solving the XOR problem. The XOR problem is related to the processing of two unequal inputs (the expected outcome is false) and two equal inputs (the expected outcome is true). This problem was resolved with the advent of the backpropagation method (Rumelhart et al. 1986). The backpropagation method enabled the neural network to train a multi-layer ANN. The application of ANNs gained popularity after 2000 with the developments in the computing power of computer and image processing algorithms.

The architecture of typical ANNs contains networks of nodes. These nodes are organized into three types of layers – an input layer, a hidden layer, and an output layer. Each layer contains several neurons. The neurons of each layer are connected to the neurons of the next layer. There is one layer of input and output, but the hidden layer can contain multiple layers. The number of hidden layer determines the processing power and computational cost of the model. The hidden layer is not visible to the user and therefore this model is referred to as a black-box model. The basic topology of an ANN is shown in Figure 7.1.

The architecture of neural networks can be of different types depending upon the presence of layers and feedback loops. The following are the three main variants of neural networks:

A. Single-layer feed-forward network: In this type of network, the neuron of the input layer is connected to the neuron of the output layer. Since there is no computation is done on the neurons of the input layer, this is called a single-layer feed-forward network.

B. Multi-layer feed-forward network: This is a typical neural network containing all three layers – input, hidden, and output.

C. Recurrent neural network: In such a network, neurons of one layer are connected to other neurons of the same layer via a feedback loop.

The geometry of interconnection among different neurons is integral to the functioning of ANN. One neuron of each layer in the network receives signals from many other neurons from different layers. The connections between two neurons are called edges. Each edge is associated with some numerical weight. Each neuron performs two functions. It integrates all the input signals and sends the integrated signal in the form of activation functions.

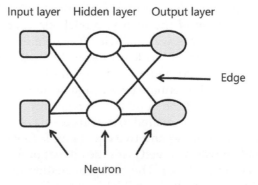

FIGURE 7.1
A typical topology of ANN containing 2 input neurons, 2 hidden layer neurons, and 2 output neurons (2-2-2).

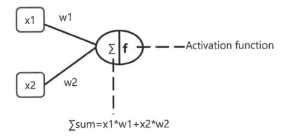

FIGURE 7.2
The neuron, edge, and summation/activation function.

The next neuron receives the activation function as an input. This is shown in Figure 7.2. The activation functions contain two factors – summation of the input signal and a factor called signal bias. This is shown as mathematical Equations (7.1) and (7.2) (Tarafdar et al. 2021). The weighting functions and signal bias factors are unknown and estimated using the experimental training data set.

$$Y_j = f^y \left(\sum_{i=1}^{n} I_i * w_{ij} + b_j \right) \tag{7.1}$$

where Y_j is the jth neuron in the hidden layer, I_i is the ith neuron in the input layer, n is the number of input neurons, w_{ij} is the weight from the ith neuron in the input neuron to the jth neuron in the hidden layer and b_j is the bias factor, and f^y is the activation function.

$$O_K = f^o \left(\sum_{j=1}^{m} Y_j * w_{jk} + b_k \right) \tag{7.2}$$

where O_k is the kth neuron in the output layer, Y_j is the jth neuron in the hidden layer, m is the number of hidden neurons, w_{jk} is the weight from the jth neuron in the hidden neuron to the kth neuron in the output layer and b_k is the bias factor, and f^o is the activation function.

The activation function can be of different types. The common activation functions are given below:

1. Linear activation function: The activation function is linear in nature. For example,

$$f(x) = x + b$$

where x is a weighted sum and b is the bias factor.

2. Heaviside step function: The activation function is in the form of step function. For example,

$$f(x)\begin{cases}1\ x \geq a \\ 0\ x < a\end{cases}$$

where x is a weighted sum and a is a threshold value.

3. Sigmoidal function: This function is of the following form:

$$f(x) = \frac{1}{1 - e^{-x}}$$

where x is a weighted sum and e is an exponential function

A linear function is usually a common form of activation function in the output layer. In the hidden layer, a sigmoidal function is a common form of activation function in biological processes like enzyme production.

7.3.1 ANN as an Optimizer

ANN model contains two key parameters – weights and signal bias. There could be millions of these parameters depending upon the complexity of networks. The training of an ANN model requires the estimation of these parameters using the training data sets. It was difficult to train the perceptron model due to the presence of the step function which is a non-differentiable function. A sigmoidal function has replaced the step function which is a differential function. This became possible with the advent of the backpropagation method.

The weights and signal bias parameters are estimated using the optimizer. An optimizer is an algorithm containing a cost or optimization function and the strategy is to minimize the cost function. The backpropagation made it possible to minimize the cost function by gradient descent. The cost function is a mathematical function that measures the deviation between the model-predicted value and the experimental value.

The presence of a large number of parameters in the ANNs makes the optimization problem highly non-convex. A non-convex optimization problem is characterized by the presence of many local valleys (local minima) and very narrow global minima. This makes the training of deeper networks difficult and requires appropriate techniques to solve the optimization problem. There are many techniques to perform the minimization of the objective function. The gradient descent method is the most common strategy to minimize the objective function. It is a derivative-based method. It is an iterative optimization technique. In this technique, a derivative (slope) is calculated at a random point. This is followed by steps proportional to the negative of the gradient of the optimization function. If the slope is large, then the bigger step is taken. If the slope is smaller, then smaller steps are taken. As it is difficult to get the global minima, a number of techniques have been proposed. These include stochastic gradient descent (SGD) (Robbins and Monro 1951), network for adversary generation (NAG), and adaptive moment estimation (ADAM) (Kingma and Ba 2015). In the stochastic descent gradient method, the optimization is done interactively by changing the objective function and training data set in each iteration. The underlying aim of all these algorithms is to seek global minima. The overfitting of the ANN model is one of the common issues. This issue is addressed by switching off some neurons during the optimization process. This method is called drop-out (Srivastava et al. 2014).

ANNs can be used for the optimization of process parameters for enzyme production. ANN is very flexible in using inputs. The inputs to the ANN model are called features and can be anything. This feature of ANN enables the model domain independent. The process parameters are used as an input to the ANN model. The yield of an enzyme is the output of the model. The optimization using the ANN model can be formulated as a supervised learning problem. An optimization function is defined which is the difference between the model output yield of the enzyme and the experimental output of the enzyme. The experimental data is divided into three categories – training, test, and validation. The model is trained by minimizing the objective function. A termination criterion is selected before the optimization process. The terminal criteria define how much minimization of the cost function is acceptable. The termination criteria are decided based on the noise in the experimental data and the accepted variance between the model and experimental output. The sets of input values (process parameters) are the optimal values for the desired yield. The workflow for setting up the optimization of process parameters is shown in Figure 7.3.

7.3.2 Example of ANN in Optimization of Process Parameters for Various Classes of Enzymes

ANN can be used for modeling bioprocesses and prediction of fermentation variables (Thibault et al. 1990). The main advantage of using ANN in the optimization process is that it can take any input and function (Cybenko 1989). These methods are suitable for process optimization (del Rio-Chanona et al. 2016; Mondal et al. 2021).

Amylases are starch-converting enzymes and are widely used in many industries, including food, fermentation, textile, detergent, paper, and pharmaceutical industries. The amylase industry is one of the largest enzyme industries (Bui et al. 2020). Amylases can be produced from different sources – animals, plants, and microorganisms. However, microorganisms

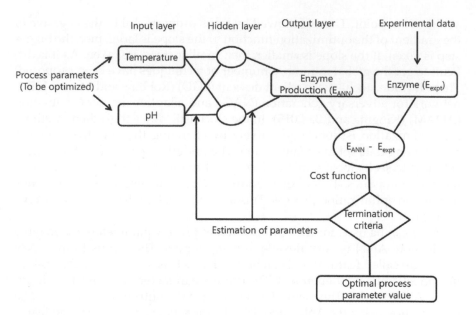

FIGURE 7.3
Workflow of using ANN as an optimizer.

(bacteria and fungi) are the preferred sources due to the ease of fermentation either solid substrate fermentation or submerged fermentation and purification. The optimal conditions for the process parameters are dependent on industrial use. Amylase should be stable at low pH for the starch industry but for the detergent industry, stability at high pH is required.

Prabhakar and Madhu (2010) used the ANN for the optimization of six inputs for the production of α-amylase from banana peel using *Bacillus subtilis* MTCC 441. The inputs included substrate content (banana peel), peptone, pH, incubation time, and inoculum size using a three-layered ANN with configuration (6-39-1), i.e., 6 input neurons, 39 hidden layer neurons, and 1 output layer. They used data set from 36 experimental runs divided into training data set (36 experimental runs) and validation data set (10 experimental runs). They obtained a good correlation coefficient (R) of 0.986 for the relation between the experimental and ANN model-predicted enzyme activity showing the generalization capability of the ANN.

Mishra et al. (2016) employed a simple three-layered feed-forward neural network in the optimum operating conditions for the production of α-amylase from *Gliomastix indicus*. They used solid surface fermentation using mustard oil cake as a solid substrate. In the optimization setting using ANN, they fixed desired activity of α-amylase and identified the optimal conditions for moisture content, incubation time, incubation temperature, pH, and concentration of sodium chloride and ammonium nitrate (source of nitrogen source). The low absolute error of 2.37 with the test data set showed that the ANN model predictions for the optimal operating condition can be generalized.

A synthetic substrate is very expensive for the production of α-amylase. Ousaadi et al. (2021) used citrus fruit feel as a substrate in submerged fermentation for the production of α-amylase by *Streptomyces* sp. They optimized four process parameters – substrate content (orange peel), sodium chloride powder, inoculum size, and pH – as an input to the ANN. They identified the optimal configuration of ANN for the optimization to be 4 input neurons, 3 hidden layer neurons, and 1 output neuron. They achieved a 4-fold increase in the α-amylase activity using the optimal conditions for the input parameters.

The rate functions in the ANN are the important determinant in the application of the ANN in the optimization of process parameters for the production of enzymes. Glucoamylases are one of the important enzymes which act on partially processed starch. These enzymes are used in the food and beverages industry. Chang et al. (2006) established the best rate functions in the ANN for the production of glucoamylase by *Monascus anka*. They used three-layered ANN using two inputs – dissolved oxygen and temperature to identify the log-sigmoidal function for the hidden layer and linear function for the output layer to be optimum for the optimization of the production of glucoamylase.

Inulinases are hydrolyzing enzymes that primarily work on inulin, levan, and sucrose. These enzymes have many applications in the food industry including the production of high fructose syrup. The selection of the kinetic model to identify the roles of different variables on inulinase production. Mazutti et al. (2009) employed the ANN to study the roles of five common variables – time of cultivation, pH, initial molasses concentration, initial corn steep liquor (CSL) concentration, and initial total reducing sugar (TRS) concentration – on the production of inulinase using batch cultivation of the yeast *Kluyveromyces marxianus* NRRL Y-7571. Using the experimental results, they established the configuration of the ANN with 5 input neurons, 10 hidden neurons, and 3 output neurons (5-10-3). They compared the ANN prediction with different kinetic models and found that the ANN prediction was better which can be used for the optimization of input variables for the production of inulinase.

Cellulase enzymes degrade cellulose and some polysaccharides. These enzymes have many applications in the textile industry. ANN can be used to improve cellulase production by optimization of culture conditions and the concentration of substrate. Lakshmi et al. (2020) used the ANN for selecting and validating the optimum values of four important factors – pH, temperature, substrate concentration, and incubation time – for the production of cellulase by *Streptomyces durhamensis* vs15. They developed an improved strain of *S. durhamensis* using mutagenesis with an increase of cellulase by

1.87-fold. ANN was used to find the optimum values of the important factors. They used three-layered ANN with 4 input neurons, 7 hidden neurons, and 1 output neuron. ANN can also be used to study specific factors of interest in the production of cellulase enzyme. Singh et al. (2008) used ANN to validate the effect of moisture content on cellulase production under solid surface fermentation using bagasse as an inducer by *Trichoderma reesei*. They compared the ANN prediction of the effect of moisture content on cellulase production with the experimental data and found the ANN as a suitable predictive tool to approximate multivariate functions.

Chitinase degrades chitin which is the second most abundant natural substance after cellulase. Chitin is present in insects, crustacean shells, and bacteria. Chitinases can be used as pesticides, and they have many applications in the clinical and pharmaceutical industry (Rathore and Gupta 2015). Suryawanshi et al. (2020) optimized incubation time, pH, and inoculum size on the production of chitinase using the fungus *Thermomyces lanuginosus* MTCC 9331. They used ANN configuration with 3 input neurons, 5 hidden neurons, and 1 output neuron. They obtained the chitinase yield of 128.53 U/L with 5 days of incubation time, pH of 5.6, and 4% inoculum. They obtained a very good correlation (>0.9 R^2 value) of ANN predicted value with experimental data. Ismail et al. (2019) used deep neural networks for the optimization of process parameters for the production of exochitinase by the fungus *Alternaria* sp. strain using solid surface fermentation. A deep neural network is a variant of ANN with higher hidden layers. They used seven process parameters for the optimization process for the production of exochitinase. They obtained an 8.5-fold increase (2.3–28.931 IU/g dry substrate) in the enzyme production using optimization by a deep neural network.

ANN can be integrated with other optimizing techniques such as genetic algorithms. The other techniques can be used to reduce the input space or enhance the optimizing routine. The combination of ANN with the other optimizing techniques increases the power of ANN which helps in addressing more complex data sets. The combination has been implemented successfully in optimizing the process parameters for enzyme production.

Alkaline protease is an important protease and is used in many industries including detergent, leather, silver recovery, pharmaceutical, waste treatment, and food industry (Sharma et al. 2019). Subba Rao et al. (2008) used ANN for the optimization of six process parameters for the optimization of alkaline protease production by *Bacillus circulans*. They used 6 input neurons, 13 hidden layer neurons, and 1 output neuron topology (6-13-1) of ANN for the optimization. The six inputs selected for optimization were glucose, soybean meal, temperature, volume, and inoculum size. They had used genetic algorithms to refine the search for the input parameters for ANN. This hybrid methodology yielded a significant increase (>2.5-fold) in enzyme production. Such an approach can be used to scale up the production scale of the enzyme production.

Lipases are hydrolyzing enzymes that convert triglycerides into their constituents such as diglycerides, monoglycerides, glycerol, and fatty acids. These are used in many industries such as food processing, detergent formulation, pharmaceutical, textile, cosmetic, and paper industry (Houde et al. 2004). Lipases constitute the third-largest enzyme group in terms of sales volume after protease and amylase (Hasan et al. 2006). Aziz et al. (2020) employed the ANN-Genetic algorithm hybrid methodology in optimizing eleven process parameters for the production of alkaline lipase production by *Nocardiopsis* sp. strain NRC/WN5. They used 11 input neurons, 100 hidden layer neurons, and 1 output neuron to optimize the process parameters involving media composition and operating condition.

Laccase catalyzes the oxidative degradation of phenolic compounds and is used in many biotechnology industries. Chiranjeevi et al. (2014) employed an ANN-Genetic algorithm hybrid methodology to optimize six process parameters for the production of laccase using solid-state fermentation by *Pleurotus ostreatus*. They achieved 2-fold increased laccase production by optimizing six parameters involving glucose concentration, peptone concentration, copper sulfate, temperature, and inoculum concentration.

7.4 Description and Application Response Surface Methodology for Process Optimization

RSM was proposed by Box and Wilson as an optimization method for chemical production processes (Box and Wilson 1951). The Box and Wilson method is also known as RSM. The motivation for the development of this optimization method was to maximize the yield which is analyzed as a surface that depends on different input values. RSM is associated with the design of the experiment. This method employs different ways to identify the optimum value of the input variables. The visualization of a surface plot of response (the product yield) helps in the identification of the sensitivity of a variable. The steepest ascent indicates whether an increase in the variable will increase the product yield. The response surface plot helps to identify the direction of the steepest ascent toward the peak of the response surface. Experiments are performed to get the peak of the response surface. The common design of the experiment is the two-level factorial design. One of the important findings using Box and Wilson's methodology was the realization that the steepest ascent may not be sufficient for maximizing the yield on the industrial production scale because of the interaction among the different inputs. So, they suggested the estimation of the curved surface for obtaining the desired surface during the optimization process. Box and Wilson introduced a composite central design (CCD) of the experiment which helps to estimate curvatures and interaction effects. The CCD is like the partial factorial design with factors at five levels. The other important strategy to achieve the optimum values of the input variable is to identify the stationary phase on the plot. The stationary region on the surface plot indicates the robustness of the optimization process. A path along the steepest descent indicates the direction of the presence of a peak of the response surface. A canonical analysis is done to locate the stationary point on the response surface. The stationary point can be classified as an optimum point, saddle, or just as a ridge. Canonical analysis requires a factorial design either full or partial to estimate the yield surface (Chen 1994; Teruel et al. 1997).

RSM is used for the optimization of process parameters for enzyme production. The optimization can be done using the following steps:

Step 1: The process parameters are identified which are expected to influence the enzyme production and increase the yield of an enzyme. The common process parameters include media composition, operating conditions (pH, time, temperature), inoculum size, and agitation speed.

Step 2: Using the statistical design, the parameters are selected for the optimization process.

Step 3: The DOE is selected using selected process parameters. There are many software packages are available to perform the design of the experiment. Some examples of software programs include Minitab and Design-Expert.

Step 4: The optimization process is performed. This step involves the development of multivariable regression models. These models relate the dependent variable (enzyme yield) to the independent variables (process parameters). The experimental results are analyzed using RSM.

RSM is suitable for the optimization of process parameters for enzyme production as this methodology could evaluate the effect of the process parameters on the response (enzyme yield). Rath et al. (2022) optimized the operating conditions (pH, temperature, and hydrogen peroxide) for the production of lignin peroxidase from *Bacillus mycoides*. Using RSM methodology, they obtained a 1.76-fold increase in the activity of lignin peroxidase compared to un-optimized conditions of operating condition parameters. Vaidya et al. (2003) obtained a 1.47-fold increase in the production of chitinase from *Alcaligenes xylosoxydans* by optimizing medium components using the RSM method. Handa et al. (2016) employed RSM methods to increase the production of pectinase by *Rhizopus* sp. They used a central composite design to degenerate the data under solid-state fermentation.

RSM methodology helped in establishing the use of statistical methods in the chemical industries including the fermentation methods. These methods have been widely used in the optimization of the production of many bio-chemical compounds (Gouveia et al. 2001; Gouveia et al. 2001; Singh and Tripathi 2008; Praveen et al. 2008; Banga and Tripathi 2009).

7.5 Comparison of ANN and RSM Methodology

RSM is a statistical method associated with the experimental design. RSM provides a visual surface plot to evaluate the effect of a single or combination of variables on the production of an enzyme (response). This enables the ease of using the method for the optimization of the process parameters. ANNs are machine learning models developed to address complex data sets. ANN models require a training process where the parameters of the model are esti-mated using the training data sets. The topology of ANN is the critical step in the training process. Therefore, ANN is a more time-consuming technique compared to RSM.

RSM involves the development of a multivariate regression model. These models can be of first-order or second-order depending upon the presence of interaction among variables. However, the second-order polynomial equa-tions result in less optimal values (Baishan et al. 2003). ANN provides several fitness functions that can be adapted as per the complexity present in the data set. This makes ANN be suitable for addressing the non-linearity in the data set.

ANN models do not limit the number of inputs in the optimization pro-cess. Anything can be used as an input to the ANN models. Further, these models are domain-independent. So, these models can be adapted to many applications. In contrast, the application of the RSM model in the optimiza-tion process gets difficult with many variables.

Several studies have been published where the performance of RSM and ANN methods for the optimization of process parameters in the production of enzymes were compared. ANN provides a better fit of the predicted optimal parameters values and enzyme yield with experimental data compared with RSM. Dutta et al. (2004) compared the normalized mean square error (MSE) obtained in the optimization of protease production by Pseudomonas sp and found 0.05% MSE with the ANN model whereas 0.1% MSE with the RSM model. Similarly, Sahu et al. (2019) found a better prediction ability of ANN compared to RSM in the optimization of cholesterol oxidase production by *Streptomyces olivaceus*. They compared Pearson correlation ANN 0.98 > RSM 0.95), regression coefficient (ANN 0.96 > RSM 0.90), and absolute average deviation (ANN 3.46% < RSM 9.87%). Ismail et al. (2019) also reported a significantly higher regression coefficient (ANN 0.99 > RSM 0.76) for ANN for the optimization of the production of exochitinase by *Alternaria* sp. Desai et al. (2008) optimized media composition for the production of scleroglucan and found a lower prediction error for the ANN-GA hybrid model compared to the RSM model (2% for ANN-GA vs 8% for RSM). The better fit of the predicted results with ANN models shows the superiority of the neural networks model in addressing the non-linear relationship between the process parameters and enzyme production.

ANN models show a better yield of enzymes compared with RSM models. Aziz et al. (2020) obtained a 1.24-fold higher average yield of thermo-alkaline lipase by *Nocardiopsis* spp. using optimal fermentation conditions by a neural network coupled with GA compared with the RSM method. Similarly, Prabhakar and Madhu (2010) reported a higher amylase production by *B. subtilis* using ANN optimized parameters compared to RSM with a central composite design. In their work, ANN optimized incubation time was half (10.12 h vs 19.6 h) compared to the RSM method showing that ANN could predict more efficient conditions for the production of enzymes.

7.6 Conclusions

The increase in enzyme production at a low cost is the need for the growing enzyme industry. This can be achieved by allocating the optimal resources for the production process. The optimization of various process parameters involved in media composition and operating conditions require robust and evolving techniques. ANN suits these requirements as it provides flexibility in using inputs of diverse types. ANN is domain-independent which helps in adapting the progress made in other fields for the optimization of process parameters for enzyme production. In this chapter, the key strategies in the optimization process are discussed. The real data generated during the production process is complex and may not be addressed by a single strategy. ANN is flexible in incorporating changes required as per the requirements.

In this chapter, the applications of ANN in the optimization of process parameters for various classes of enzymes are presented using the published results. The underlying approaches of ANN for these different classes of enzymes are the same. However, the topology of ANN might be different depending on the data. This requires a skilled person to perform the optimization process.

In this chapter, ANN methods are compared with RSM. The comparison helps to understand the difference between the two methods. ANN models show superiority in prediction accuracy and handling of the non-linear relationship between process parameters and enzyme production. The published performance of ANN compared to RSM shows that ANN can be a good alternative for the optimization of process parameters of enzyme production.

References

Aziz, Mohamed M. Abdel, Eman W. Elgammal, and Roba G. Ghitas 2020. Comparative study on modeling by neural networks and response surface methodology for better prediction and optimization of fermentation parameters: Application on thermo-alkaline lipase production by Nocardiopsis sp. strain NRC/WN5.25. *Biocatalysis and Agricultural Biotechnology* 25: 101619.

Baishan, F., C. Hongwen, X. Xiaolan, W. Ning, and H. Zongding 2003. Using genetic algorithms coupling neural networks in a study of xylitol production: Medium optimization. *Process Biochemistry* 38: 979–985.

Balaman, Ş. Y. 2019. Modeling and Optimization Approaches in Design and Management of Biomass-Based Production Chains. In *Decision-Making for Biomass-Based Production Chains*, pp. 185–236.

Banga, J., and C. K. M. Tripathi 2009. Response surface methodology for optimization of medium components in submerged culture of Aspergillus flavus for enhanced heparinase production. *Letters in Applied Microbiology* 49: 204–209.

Box, G. E. and K. B. Wilson 1951. On the experimental attainment of optimum conditions. *J Royal Statistical Society: Series B (Methodological)*, 13, no. 1: 1–38.

Bui, Alexander T., Barbara A. Williams, Emily C. Hoedt, Mark Morrison, Deirdre Mikkelsen, and Michael J. Gidley 2020. High amylose wheat starch structures display unique fermentability characteristics, microbial community shifts and enzyme degradation profiles. *Food & Function* 11, no. 6:5635–5646.

Chang, Jyh-Shyong, Jinn-Tsair Lee, and Audrey-Chingzu Chang 2006. Neural-network rate-function modeling of submerged cultivation of Monascus anka. *Biochemical Engineering Journal* 32, no. 2: 119–126.

Chen, Hung-Chang 1994. Response-surface methodology for optimizing citric acid fermentation by Aspergillus foetidus. *Process Biochemistry* 29, no. 5: 399–405.

Chiranjeevi, Potu Venkata, Moses Rajasekara Pandian, and Sathish Thadikamala 2014 Integration of artificial neural network modeling and genetic algorithm approach for enrichment of laccase production in solid state fermentation by Pleurotus ostreatus. *BioResources* 9, no. 2: 2459–2470.

Cybenko, George 1989. Approximation by superpositions of a sigmoidal function. *Mathematics of Control, Signals, and Systems* 2, no. 4: 303–314.

Dantzig, George B., and Mukund N. Thapa 1997 *Linear Programming. 1, Introduction {Springer Series in Operations Research}*. Springer-Verlag, New York.

del Rio-Chanona, Ehecatl, Antonio, Emmanuel Manirafasha, Dongda Zhang, Qian Yue, and Keju Jing 2016. Dynamic modeling and optimization of cyanobacterial C-phycocyanin production process by artificial neural network. *Algal Research* 13: 7–15.

Desai, Kiran M., Shrikant A. Survase, Parag S. Saudagar, S. S. Lele, and Rekha S. Singhal 2008 Comparison of artificial neural network (ANN) and response surface methodology (RSM) in fermentation media optimization: Case study of fermentative production of scleroglucan. *Biochemical Engineering Journal* 41, no. 3: 266–273.

Dutta, Jayati Ray, Pranab Kumar Dutta, and Rintu Banerjee 2004 Optimization of culture parameters for extracellular protease production from a newly isolated Pseudomonas sp. using response surface and artificial neural network models. *Process Biochemistry* 39, no. 12: 2193–2198.

Gluss, Brian 1961. An introduction to dynamic programming. *Journal of the Staple Inn Actuarial Society* 16, no. 4: 261–274.

Gouveia, E. R., A. Baptista-Neto, A. C. Badino, Jr., and C. O. Hokka 2001. Optimization of medium composition for clavulanic acid production by Streptomyces clavuligerus. *Biotechnology Letters* 23: 157–161.

Handa, Shweta, Nivedita Sharma, and Shruti Pathania 2016. Multiple parameter optimization for maximization of pectinase production by Rhizopus sp. C4 under solid state fermentation. *Fermentation* 2, no. 2: 10.

Hasan, F., A.A. Shah, and A. Hameed 2006. Industrial applications of microbial lipases. *Enzyme and Microbial Technology* 39, no. 2: 235–251.

Houde, Alain, Ali Kademi, and Danielle Leblanc 2004. Lipases and their industrial applications. *Applied Biochemistry and Biotechnology* 118, no. 1: 155–170.

Ilkova, Tatiana, and Mitko Petrov 2008. Dynamic and neuro-dynamic optimization of a fed-batch fermentation process. In *International Conference on Artificial Intelligence: Methodology, Systems, and Applications*, pp. 365–369. Springer, Berlin, Heidelberg.

Ismail, Shaymaa A., Ahmed Serwa, Amira Abood, Bahgat Fayed, Siham A. Ismail, and Amal M. Hashem 2019. A study of the use of deep artificial neural network in the optimization of the production of antifungal exochitinase compared with the response surface methodology. *Jordan Journal of Biological Sciences* 12, no. 5: 541–551.

Kennedy, Max, and Donal Krouse 1999. Strategies for improving fermentation medium performance: A review. *Journal of Industrial Microbiology and Biotechnology* 23, no. 6: 456–475.

Kingma, D. P., and Ba, J. L. 2015. Adam: A method for stochastic optimization. In *International Conference on Learning Representations*, pp. 1–13.

Klein-Marcuschamer, Daniel, Piotr Oleskowicz-Popiel, Blake A. Simmons, and Harvey W. Blanch 2012. The challenge of enzyme cost in the production of lignocellulosic biofuels. *Biotechnology and Bioengineering* 109, no. 4: 1083–1087.

Lakshmi, Ega Soujanya, Manda Rama Narasinga Rao, and Muddada Sudhamani 2020. Response surface methodology-artificial neural network based optimization and strain improvement of cellulase production by Streptomyces sp. *Bioscience Journal*: 1390–1402. https://seer.ufu.br/index.php/biosciencejournal/article/view/48006/28992.

Luenberger, D.G. 1973. *Introduction to Linear and Non-Linear Programming*. Addison-Wesley Publishing Company, Reading, MA.

Mandenius, Carl-Fredrik, and Anders Brundin 2008. Bioprocess optimization using design-of-experiments methodology. *Biotechnology Progress* 24, no. 6: 1191–1203.

Mazutti, Marcio A., Marcos L. Corazza, Maria Isabel Rodrigues, Fernanda C. Corazza, and Helen Treichel 2009. Inulinase production in a batch bioreactor using agro-industrial residues as the substrate: Experimental data and modeling. *Bioprocess and Biosystems Engineering* 32, no. 1: 85–95.

McCulloch, W. S., and W. Pitts 1943. A logical calculus of the ideas immanent in nervous activity. *The Bulletin of Mathematical Biophysics*, 5(4): 115–133.

Minski, M. L., and S. A. Papert 1969. *Perceptrons: An Introduction to Computational Geometry.* MIT Press, Cambridge, MA.

Mishra, Santosh Kumar, Shashi Kumar, Surendra Kumar, and Ravi Kant Singh 2016 Optimization of process parameters for-amylase production using Artificial Neural Network (ANN) on agricultural wastes. *Current Trends in Biotechnology and Pharmacy* 10, no. 3: 248–260.

Möller, Philip, Xiaochen Liu, Stefan Schuster, and Daniel Boley 2018, Linear programming model can explain respiration of fermentation products *PLoS One* 13, no. 2: e0191803.

Mondal, Payel, Anup Kumar Sadhukhan, Amit Ganguly, and Parthapratim Gupta 2021 Optimization of process parameters for bio-enzymatic and enzymatic saccharification of waste broken rice for ethanol production using response surface methodology and artificial neural network–genetic algorithm. 3 *Biotech* 11, no. 1: 1–18.

Ousaadi, Mouna Imene, Fateh Merouane, Mohammed Berkani, Fares Almomani, Yasser Vasseghian, and Mahmoud Kitouni. 2021. Valorization and optimization of agro-industrial orange waste for the production of enzyme by halophilic Streptomyces sp. *Environmental Research* 201: 111494.

Prabhakar, A., and G. M. Madhu 2010 Statistical optimization and neural modeling of amylase production from banana peel using Bacillus subtilis MTCC 441. *International Journal of Food Engineering* 6, no. 4. doi: 10.2202/1556-3758.1980

Praveen, V., D. Tripathi, C. K. M. Tripathi, and V. Bihari 2008. Nutritional regulation of actinomycin-D production by a new isolate of Streptomyces sindenensis using statistical methods. *Indian Journal of Experimental Biology* 46: 139–144.

Rath, Subhashree, Manish Paul, Hemanta Kumar Behera, and Hrudayanath Thatoi 2022. Response surface methodology mediated optimization of Lignin peroxidase from Bacillus mycoides isolated from Simlipal Biosphere Reserve, Odisha, India. *Journal, Genetic Engineering & Biotechnology* 20, no. 1: 1–20.

Rathore, Abhishek Singh, and Rinkoo D. Gupta.2015 Chitinases from bacteria to human: Properties, applications, and future perspectives. *Enzyme Research* https://www.ncbi.nlm.nih.gov/pmc/articles/PMC4668315/

Robbins, Herbert, and Sutton Monro.1951. A Stochastic Approximation Method. In *The Annals of Mathematical Statistics*: 400–407.

Rosenblatt, F. 1962. *Principles of Neurodynamics, Science Editions*, New York.

Rumelhart, D. E., G. E. Hinton, and R. J. Williams 1986. Learning representations by back-propagating errors. *Nature*, 323(6088), 533.

Sahu, Shraddha, Shailendra Singh Shera, and Rathindra Mohan Banik 2019. Optimization of process parameters for cholesterol oxidase production by Streptomyces Olivaceus MTCC 6820. *The Open Biotechnology Journal*, 13, no. 1: 47–58.

Sharma, Mayuri, Yogesh Gat, Shalini Arya, Vikas Kumar, Anil Panghal, and Ashwani Kumar 2019 A review on microbial alkaline protease: An essential tool for various industrial approaches *Industrial Biotechnology* 15, no. 2: 69–78.

Singh, Aruna, Divya Tatewar, P. N. Shastri, and S. L. Pandharipande 2008. Application of ANN for prediction of cellulase and xylanase production by Trichoderma reesei under SSF condition. *Indian Journal of Chemical Technology* 15: 53–58.

Singh, Rajendra, Manoj Kumar, Anshumali Mittal, and Praveen Kumar Mehta. 2016. Microbial enzymes: Industrial progress in 21st Century. *3 Biotech* 6, no. 2: 1–15.

Singh, V., and C. Tripathi 2008. Production and statistical optimization of anovel olivanic acid by Streptomyces olivaceus MTCC 6820. *Process Biochemistry*, 43, no. 11: 1313–1317.

Singh, Vineeta, Shafiul Haque, Ram Niwas, Akansha Srivastava, Mukesh Pasupuleti, and C.K.M. Tripathi 2017. Strategies for fermentation medium optimization: An in-depth review. *Frontiers in Microbiology* 7: 2087.

Srivastava, N., G. Hinton, A. Krizhevsky, I. Sutskever, and R. Salakhutdinov 2014. Dropout: A simple way to prevent neural networks from overfitting. *The Journal of Machine Learning Research*, 15, no. 1: 1929–1958.

Subba Rao, Ch, T. Sathish, M. Mahalaxmi, G. Suvarna Laxmi, R. Sreenivas Rao, and R. S. Prakasham 2008. Modelling and optimization of fermentation factors for enhancement of alkaline protease production by isolated Bacillus circulans using feed-forward neural network and genetic algorithm. *Journal of Applied Microbiology* 104, no. 3: 889–898.

Sun, Shiliang, Zehui Cao, Han Zhu, and Jing Zhao 2019. A survey of optimization methods from a machine learning perspective. *IEEE Transactions on Cybernetics* 50, no. 8: 3668–3681.

Suryawanshi, Nisha, Jyoti Sahu, Yash Moda, and J. Satya Eswari 2020. Optimization of process parameters for improved chitinase activity from Thermomyces sp. by using artificial neural network and genetic algorithm *Preparative Biochemistry & Biotechnology* 50, no. 10: 1031–1041.

Tarafdar, Ayon, Ranjna Sirohi, Vivek Kumar Gaur, Sunil Kumar, Poonam Sharma, Sunita Varjani, Hari Om Pandey et al 2021. Engineering interventions in enzyme production: Lab to industrial scale. *Bioresource Technology* 326: 124771.

Teruel, A. M. L., E. Gontier, C. Bienaime, J. Nava Saucedo, and J.-N. Barbotin 1997. Response surface analysis of chlortetracycline and tetracycline production with K-carrageenan immobilized Streptomyces aureofaciens. *Enzyme and Microbial Technology* 21: 314–320.

Thibault, Jules, Vincent Van Breusegem, and Arlette Chéruy 1990. On-line prediction of fermentation variables using neural networks. *Biotechnology and Bioengineering* 36, no. 10: 1041–1048.

Torres, Néstor V., Eberhard O. Voit, and Carlos González-Alcón 1996. Optimization of nonlinear biotechnological processes with linear programming: Application to citric acid production by Aspergillus niger *Biotechnology and Bioengineering* 49, no. 3: 247–258.

Vaidya, Rajiv, Pranav Vyas, and H. S. Chhatpar 2003. Statistical optimization of medium components for the production of chitinase by Alcaligenes xylosoxydans. *Enzyme and Microbial Technology* 33, no. 1: 92–96.

8

Advanced Evolutionary Differential Evolution and Central Composite Design: Comparative Study for Process Optimization of Chitinase Production

Nisha Suryawanshi

Government Arts and Commerce College, Sagar, India

J. Satya Eswari

National Institute of Technology, Raipur, India

CONTENTS

8.1 Introduction

Chitinase (EC 3.2.2.14), a glycosyl hydrolases (GH) enzyme, belongs to the family GH-18, well-known to hydrolyze the β-1,4 linkages of GlcNAc found in the chitin (Bhattacharya et al. 2007; Chen et al. 2017). Chitinases have been classified as endochitinases and exochitinase, where endochitinases degrade the chitin polymer by hydrolyzing the internal glycosidic bonds and creating oligomers with a small molecular mass. Chitobiosidases and N-acetylglucosaminidases are the two forms of exochitinase, which cause the NAG dimers to be released from the nonreducing end, and cleaves oligomers and dimers obtained by endochitinases and chitobiosidases into a single molecule of N-acetyl glucosamine (GlcNAc), respectively (Sahai and Manocha 1993). Chitinases from thermophilic organisms are known to have inherent thermostability; because of this property, chitinases show potential for efficient degradation of chitin into its oligosaccharides (chito-oligomers). Thermostable enzymes are known to give higher manufacturing process yields and also have the added advantage that it reduces microbial pollution caused by mesophilic species. Chitinase from thermophilic microbes which are thermostable has numerous industrial, medical, environmental, pharmaceutical, and biotechnological applications since chitinase is highly balanced at unfavorable temperatures and pH levels. The thermophilic deuteromycete *Thermomyces lanuginosus* is a broadly dispersed fungus and grows at a temperature between 20°C and 60°C, and 50°C is the optimal temperature for growth (Singh et al. 2003). A few strains of this fungus have been discovered to secrete a significant number of thermostable enzymes involving invertase, protease, trehalase, xylanase, and chitinase(Karn and Kumar 2019; Karn and Duan 2017). Chitinases are utilized in various applications, comprising waste management in agriculture as pest control and in human health care. Chitinase is also used as an antifungal agent, blended with antifungal drugs in rehabilitation for many fungal infections (Oranusi and Trinci 1985).

Chito-oligomers have also gained much attention in pharmaceutics due to their biocompatibility and nontoxicity. Chitooligomers and GlcNAc obtained from Chitin are used in various applications, comprising pharmaceuticals and the food industry. Moreover, chitooligomers are also consumed for numerous reasons, such as antimicrobial activity, antioxidant effect, and drug delivery. It is generally observed that the yield of chitinase is higher in the presence of chitin as a substrate in the production medium. Chitin is a polymer with a high molecular weight that is insoluble in water. As a result, chitin polymer consumption and degradation within the cell by microorganisms are problematic. Therefore, microbes produce enzymes that would dissolve chitin into simpler forms, which can then absorb nutrients. The favorable media components are necessary for chitinase production, which can help microbes develop faster while also increasing the development of chitin-degrading enzymes.

Furthermore, a production medium must be developed for a cost-effective process. There is a wealth of knowledge available for optimization based on one-factor-at-a-time (OFAT) and statistical methods for the formulation of the medium components to promote microorganism growth and the development of chitin-degrading enzymes (Kumar et al. 2012). However, dealing with many variables proved to be a lengthy, costly, and unmanageable process using the OFAT method (Dahiya et al. 2005).

Furthermore, this experimental approach necessitates many trials to find the optimum levels, which are inconsistent (Vaidya et al. 2003). Experiments with a statistical design will help overcome the OFAT method's limitations and maximize the medium's influencing variables (Khandelwal et al. 2007; Mak et al. 1995; Montgomery 2001; Chandrasekhar et al. 2021; Chandrasekhar et al. 2020). The application of a statistical design process has several benefits, including the fact that it is fast and accurate for nutrient shortlisting; it is simple to understand the value of media constituents at various concentrations, and It can drastically decrease the number of trials performed. The statistical approach's benefits include saving electricity, time, glassware, and chemicals (Vaidya et al. 2003). To develop the medium components, statistical models, including response surface methodology (RSM) (Roudi et al. 2020), are effective tools (Montgomery 2003; Doddapaneni et al. 2007; Alslaibi et al. 2013; Eswari et al. 2013). The simplest way to easily identify the important variable from many variables is to use experiment design in statistical techniques. One method is the Plackett–Burman design (PBD) which is widely used to screen many variables simultaneously (Plackett and Burman 1946; Yu et al. 1997). RSM has been used for the optimization of chitinase production from filamentous fungi (Wasli et al. 2009; Kumar et al. 2012; Dhillon et al. 2011; Kumar et al. 2017; Liu et al. 2013). Despite the limitations of well-defined threshold aspects in RSM, complementary optimization approaches to find the actual goals are encouraged. The biological demonstration, which can give explanations to a wide range of complicated normalization challenges, is encouraging rapid advances in the field of computational approaches. Biologically encouraged evolutionary algorithms cover various nature-inspired approaches, comprising genetic algorithms and evolution strategies, which are some of the other possibilities for conventional optimization approaches.

Storn and Price (1997) suggested an evolutionary computation-based approach *viz.* differential evolution (DE) that is remarkably easy, prompt, and vigorous for resolving large standardization-related complications. The probability is more to finding the real optimal of a complication by defeating the drawbacks of other evolutionary approaches: elevated computational rate and time-consuming merging cost (Bhattacharya et al. 2011). This research aims to optimize variables to formulate a fermentation medium that will produce the chitinase enzyme from *T. lanuginosus* by applying PBD and

RSM. The regression model developed by RSM was subsequently optimized by the advanced evolutionary DE approach. To the extent of the author's belief, this is the first time RSM and DE have been used to optimize chitinase.

8.2 Methods and Materials

8.2.1 Microorganism and Growth Condition

T. lanuginosus (MTCC 9331) lyophilized culture was obtained from MTCC Institute of Microbial Technology, Chandigarh. The organism was grown for 7 d at 50°C on slants of saboured dextrose agar (SDA), stored at 4°C, and subcultured once a month to use later.

8.2.2 Substrate Preparation (Colloidal Chitin)

The procedure for making colloidal chitin was obtained from literature (Roberts and Selitrennikoff 1988; Suryawanshi and Jujjawarapu 2020). Shrimp shell chitin flakes (5 g) were dissolved in 100 mL concentrated HCl and stirred for 2 h. The mixture was added to pre-chilled distilled water and continuously stirred for 2 h. Following gentle mixing, the mixture was centrifuged at 10,000 rpm for 10 min at 4°C. To achieve a neutral pH, The colloidal chitin was cleaned in distilled water many times. Further, the colloidal chitin was heat sterilized at 121°C for 15 min and kept at 4°C for further use. Chitin (5 g) was transformed into 15 g of colloidal chitin (wet weight).

8.2.3 Production of Chitinase

To produce the chitinase enzyme 100 mL of production media containing 10 g/L colloidal chitin, 0.87 g/L, K_2HPO_4, 0.68 g/L KH_2PO_4, 0.2g/L KCl, 1.0 g/L NH_4Cl, 0.2 g/L $MgSO_4$. $7H_2O$, and 4.0 g/L yeast extract (pH 6.5) (Zhang 2014) was poured into an Erlenmeyer flask with a capacity of 250 mL and a 5% seed culture was inoculated into the medium. After that, the culture medium was incubated for 6 d at 50°C and 150 rpm. After 6 d of incubation, the medium was centrifuged for 10 min (10,000 rpm, 4°C). Chitinase activity was calculated using the supernatant as a crude enzyme (Suryawanshi and Jujjawarapu 2020).

8.2.4 Chitinase Enzyme Activity Assay

The chitinase activity was carried out on colloidal chitin as a substrate (Purushotham et al. 2012). The substrate colloidal chitin (1%) was formed in a 50 mM phosphate buffer of pH 6.5. For the reaction, 500 µL of 1% substrate colloidal chitin and 500 µL of culture supernatant were taken. The reaction mixture was kept at 50°C for 1 h in a shaking incubator followed by a water bath for 10 min at 100°C. After that, the resulting mixture was centrifuged for 10 min at 10,000 rpm. The supernatant was then analyzed to estimate the concentration of the N-acetyl-glucosamine using HPLC with Aminex HPX-87H column. The amount of chitinase used to liberate 1 micromole of NAG under reaction conditions is determined as 1 unit of chitinase activity.

8.2.5 Statistical Optimization of Medium Components

8.2.5.1 Plackett–Burman Design

With 19 variables, including four dummy variables, the PBD was used to obtain the most significant factors influencing the chitinase activity (Plackett and Burman 1946). Low-level (−1) and high-level (+1) analyses were conducted on the variables (Table 8.1). The evaluation of chitinase activity was done with 20 experiments. The experimental design of Plackett–Burman (Table 8.2) was prepared in the "Design-Expert version 12", State-Ease Inc.,

TABLE 8.1

Placket–Burman Design for Optimization of 19 Factors Each at Two Levels for the Production of Chitinase Enzyme

Factor	Name	Low Level (−1)	High Level (+1)
A	Colloidal chitin	1	10
B	Glucose	0.2	2
C	KNO_3	2	10
D	Yeast extract	0.5	5
E	KH_2PO_4	0.14	0.68
F	Na_2HPO_4	0.14	0.7
G	$MgSO_4$	0.02	0.2
H	$MnSO_4$	0.0001	0.001
J	$CaCl_2$	0.005	0.05
K	KCl	0.2	2
L	$ZnSO_4$	0.0001	0.001
M	$FeSO_4$	0.002	0.02
N	$CoCl_2$	0.0001	0.001
O	Tween 80	0.1	1
P	Triton X	0.1	1
Q	Dummy 1	−1	1
R	Dummy 2	−1	1
S	Dummy 3	−1	1
T	Dummy 4	−1	1

TABLE 8.2

Experimental Design and Results of Placket–Burman Design

Run	A:Colloidal Chitin	B:Glucose	C:KNO$_3$	D:Yeast Extract	E:KH$_2$PO$_4$	F:Na$_2$HPO$_4$	G:MgSO$_4$	H:MnSO$_4$	J:CaCl$_2$	K:KCl	L:ZnSO$_4$
	%	%	%	%	%	%	%	%	%	%	%
1	−1	1	1	−1	−1	1	1	1	1	1	1
2	−1	−1	−1	1	−1	−1	−1	−1	−1	−1	−1
3	1	1	−1	1	1	−1	1	1	−1	1	1
4	−1	1	−1	1	1	−1	1	1	−1	1	1
5	1	−1	−1	−1	−1	1	1	1	1	1	1
6	−1	−1	−1	−1	1	1	1	1	1	1	1
7	1	1	−1	1	1	1	1	1	1	1	1
8	1	−1	−1	1	1	1	1	1	1	1	1
9	−1	1	−1	−1	1	1	1	1	1	1	1
10	1	1	1	1	1	1	1	1	1	1	1
11	1	−1	1	1	1	1	1	1	1	1	1
12	−1	−1	1	1	1	1	1	1	1	1	1
13	−1	−1	1	1	1	1	1	1	1	1	1
14	1	−1	1	1	1	1	1	1	1	1	1
15	1	−1	−1	1	1	1	1	1	1	1	1
16	1	1	−1	1	1	1	1	1	1	1	1
17	−1	−1	−1	−1	1	1	1	1	1	1	1
18	−1	−1	−1	1	1	1	1	1	1	1	1
19	1	1	−1	−1	1	1	1	1	1	1	1
20	−1	1	−1	1	−1	−1	1	1	1	1	1

M:FeSO$_4$	N:CoCl$_2$	O:Tween 80	P:Triton X	Q:Dummy 1	R:Dummy 2	S:Dummy 3	T:Dummy 4	R1 (Chitinase Activity)	R2 (Biomass)
%	%	%	%	%	%	%	%	U/L	g/L
-1	1	-1	-1	-1	-1	1	1	8.04	0.21
-1	1	1	-1	1	-1	1	-1	11.95	0.83
-1	-1	1	1	-1	-1	1	1	16.54	1.07
-1	1	-1	-1	1	1	1	1	7.51	0.845
-1	-1	-1	-1	1	1	1	-1	15.27	0.775
-1	-1	-1	1	1	-1	1	1	9.30	0.885
-1	-1	1	-1	-1	1	-1	-1	24.03	1.405
-1	1	1	-1	-1	-1	-1	-1	34.18	0.475
-1	1	1	-1	-1	-1	-1	1	15.31	1.045
-1	-1	1	-1	-1	-1	-1	-1	12.55	0.84
1	1	1	-1	-1	1	1	-1	32.67	0.67
1	-1	1	-1	1	1	1	-1	28.58	1.185
1	1	-1	-1	1	1	1	-1	9.23	0.745
1	-1	-1	1	1	-1	1	1	20.36	0.55
1	1	-1	-1	-1	1	1	1	25.44	0.485
1	1	1	1	1	-1	-1	-1	22.78	0.84
1	-1	-1	-1	1	1	1	-1	11.59	0.52
1	1	-1	-1	1	1	-1	-1	14.58	0.5
1	1	-1	-1	1	1	-1	1	28.54	0.87
-1	-1	1	1	1	1	-1	1	22.70	1.045

Minneapolis, USA, statistical software with 19 variables for 20 experiments. The experiments were conducted thrice and evaluated by the same software.

8.2.5.2 The Central Composite Design and the Response Surface Methodology

The four very significant variables (yeast extract, Tween 80, glucose, and colloidal chitin) obtained from the PBD were then standardized for optimum levels by utilizing a central composite design to boost the activity of chitinase. The objective factors were evaluated with five levels, i.e., $-\alpha$, -1, 0, $+1$, $+\alpha$, $(-1 =$ low level, $+1 =$ high level, $\alpha = 2k/4$, where k is the no. of factors), 30 experiments (Table 8.3) was performed with 8 axial points, 16 factorial points, and 6 center points. The results obtained by experiments were evaluated using statistical software "Design-Expert version 12", State-Ease Inc., Minneapolis, USA. The mean value of chitinase activity was reffered to as response (R). Analysis of variance (ANOVA) was done to obtain an accurate model that reveals the importance of objective variables and determines their interaction. The regression equation described the nature of the model-

$$Y = \beta_0 + \sum_{i=1}^{3} \beta_i X_i + \sum_{i=1}^{3} \beta_{ii} X_i^2 + \sum_{ij=1}^{3} \beta_{ij} X_{ij} \qquad (8.1)$$

where $Y =$ expected response, $\beta_0 =$ balanced term, $\beta_i =$ constant for the linear terms, $\beta_{ii} =$ constant for square terms, and $\beta_{ij} =$ constant of interactive terms. The optimum levels of four significant factors were determined using the regression equation.

8.2.6 Machine Learning Approach: Differential Evolution

8.2.6.1 Differential Evolution in the Context

Kenneth and Storn (1997) introduced DE which is recognized as a far more prominent optimizer for solving complex optimization issues among metaheuristic search algorithms (MSAs) established in the last few years. DE is a community-based approach that is commonly used to tackle numerous sorts of optimization issues and is part of the evolutionary algorithm (EA) family. In contrast to other EAs which develop progeny by troubling solutions with graded variance vectors, it produces new progeny by recombining solutions within particular conditions. If the new offspring solution exceeds the present individual solution, it will be replaced (Price et al. 2006). Because its search process is governed by several automated process variables, including scaling factor and crossover rate, DE is regarded as a robust and

TABLE 8.3

Central Composite Design Model Experimental Design Matrix (n = 4)

Run	Factor 1 A: Colloidal Chitin %	Factor 2 B: Glucose %	Factor 3 C: Tween 80 %	Factor 4 D: Yeast Extract %	Response Observed R1 U/L	Response Predicted R1 U/L
1	1	0.25	0.07	6	90.51	90.19
2	1	0.25	0.03	6	86.74	80.86
3	1	0.15	0.07	6	101.82	98.54
4	0.8	0.2	0.05	5	77.93	79.57
5	1.2	0.2	0.05	5	73.73	73.75
6	1.4	0.25	0.07	6	105.12	94.38
7	1	0.25	0.03	4	75.42	76.51
8	1.2	0.2	0.05	5	73.76	73.74
9	1	0.25	0.07	4	82.18	77.29
10	1.2	0.3	0.05	5	75.52	83.87
11	1	0.15	0.03	4	72.41	77.03
12	1.2	0.2	0.05	5	73.90	73.74
13	1.4	0.15	0.07	6	103.13	97.01
14	1.2	0.2	0.01	5	82.62	83.53
15	1.2	0.1	0.05	5	84.23	87.02
16	1.2	0.2	0.09	5	87.66	97.87
17	1.4	0.15	0.03	6	84.69	83.45
18	1.6	0.2	0.05	5	78.93	88.42
19	1.4	0.25	0.07	4	92.35	89.73
20	1.4	0.15	0.03	4	86.39	81.69
21	1	0.15	0.03	6	89.44	87.04
22	1.2	0.2	0.05	5	73.35	73.74
23	1.4	0.25	0.03	6	84.62	82.98
24	1.4	0.15	0.07	4	86.94	86.70
25	1.2	0.2	0.05	5	73.82	73.74
26	1	0.15	0.07	4	83.36	79.97
27	1.4	0.25	0.03	4	89.73	86.89
28	1.2	0.2	0.05	3	74.19	75.09
29	1.2	0.2	0.05	7	79.50	89.74
30	1.2	0.2	0.05	5	73.93	73.74

straightforward algorithm. DE, like other EAs, has three ways to generate new offspring solutions: mutation, crossover, and selection. Crossover and mutation are both recognized to have a bigger effect on the search efficiency of an algorithm (Qin et al. 2009). Though a large number of MSAs based on naturalistic phenomena have been presented in recent decades, DE is still one of the most prominent MSAs employed by both academics and practitioners to solve a wide range of real-world optimization issues due to its significant competitive advantages. To begin with, DE is easier and much simpler to deploy than furthermost MSAs (Das and Suganthan 2010). This useful aspect allows operators who do not have the robust programming expertise to create minor changes to DE's coding to resolve their domain-specific challenges. Second, due to its simple structure, DE outperforms other MSAs in handling a variety of optimization problems with hard characteristics such as nonlinearity, multimodality, and inseparability. Third, since 2005, various DE forms have consistently placed in the topmost well optimizers throughout maximum Congress of Evolutionary Computation (CEC) contests (Pant et al. 2020), inferring that DE forms have the potential for solving a variety of practical problems containing competitive searching efficiency, search durability, and faster convergence. Finally, DE seems to have the most desirable property of minimum space complexity as compared to other MSAs that may potentially process well in difficult optimization issues (Das and Suganthan 2010). In other aspects, owing to its minimum space requirements, DE has an additional flexibility than certain present MSAs in solving large-scale and computationally intensive optimization tasks.

8.2.6.2 The Fundamentals of DE Algorithms

Initialization, mutation, crossover, and selection are the four phases of basic DE's algorithmic framework, as depicted in Figure 8.1. The initialization is a one-time step, whereas the following three processes are recurrent until the termination requirements are met in the DE search progression in a D-dimensional solution space.

8.2.6.2.1 Initialization

Initialization is the initial step in DE to find a universal optimal result in D-dimensional actual constraint space. The initial results for a specific multidimensional optimization task are NP actual-valued constraint vectors, where NP denotes DE's populace size.

Each i-th particular DE result can be denoted as a D-dimensional vector shown in (Equation (8.2)) during the t-th iteration:

$$X_i^t = \left(X_{i,1}, X_{i,2}, \ldots\ldots, X_{i,D} \right) \tag{8.2}$$

where $I = 1, 2, \ldots$, NP are the integers.

At $t = 0$, the starting population condition is established. During the initialization stage, the inferior and superior limit bounds of the result examination

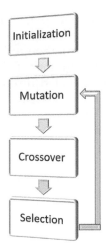

FIGURE 8.1
Subsequent stages of differential evolution.

space, denoted by (Equations (8.3) and (8.4)), respectively, can be used to generate initial candidate solutions in the following manner:

$$X_{min} = \left(X_{min,1}, X_{min,2}, \dots X_{min,D}\right) \tag{8.3}$$

$$X_{max} = \left(X_{max,1}, X_{max,2}, \dots X_{max,D}\right) \tag{8.4}$$

The j-th dimensional component of each i-th DE solution can be initiated by arbitrarily producing a value amongst the superior limit of $X_{max,\,j}$ and inferior limits of $X_{min,\,j}$ as demonstrated in (Equation (8.5)):

$$X_{i,j}^{(0)} = X_{min,j} + rand_{i,j}\left[0,1\right]\left(X_{max,j} - X_{min,j}\right) \tag{8.5}$$

where $rand_{i,\,j}[0,1]$ is an equal distribution capable of producing any actual value amongst 0 and 1.

8.2.6.2.2 Mutation

In biology, a mutation is characterized as an instantaneous change in a chromosomal gene's characteristics. The mutation is a random perturbation method used to specify decision variables in the area of evolutionary algorithms.

A mutant or donor vector indicated as Y_i^t in DE philosophy, is created via a mutation process based on a specified target vector, X_i^t (Pant et al. 2020, Neri and Tirronen 2010). The DE mutation technique is generally expressed as "DE/*/n", where n denotes the number of alteration vectors included and * denotes the objective vector evaluated throughout the mutation progression. The following equations depict the search methods for five popular DE mutation strategies.

DE/rand/1:

$$Y_i^t = X_{r_1}^t + F\left(X_{r_2}^t - X_{r_3}^t\right)$$ (8.6)

DE/rand/2:

$$Y_i^t = X_{r_1}^t + F\left(X_{r_2}^t - X_{r_3}^t\right) + F\left(X_{r_4}^t - X_{r_5}^t\right)$$ (8.7)

DE/best/1:

$$Y_i^t = X_{best}^t + F\left(X_{r_1}^t - X_{r_2}^t\right)$$ (8.8)

DE/best/2:

$$Y_i^t = X_{best}^t + F\left(X_{r_1}^t - X_{r_2}^t\right) + F\left(X_{r_3}^t - X_{r_4}^t\right)$$ (8.9)

DE/current-to-best/1:

$$Y_i^t = X_i^t + F(X_{best}^t - X_i^t + F\left(X_{r_1}^t - X_{r_2}^t\right)$$ (8.10)

where r_1 denotes the community guide of the DE result chosen as the source vector, and r_2, r_3, r_4, and r_5 denote the community index of DE results chosen at random to generate the mutant vector, where $r_1, r_2, r_3, r_4, r_5 \in [NP]$ and $r_1 \neq r_2 \neq r_3 \neq r_4 \neq r_5 \neq i$; X_{best}^t denotes that the target vector is chosen from the DE population's best individual solution; and F is a scale parameter that is used to govern the mutation progression, with a value amongst [0,1]. Selecting the right value for F is critical for achieving correct complementary of the algorithm's exploration and exploitation searches and avoiding downsides like premature convergence or poor convergence pace.

8.2.6.2.3 Crossover

In this step, the transformed and objective vectors crossover their elements in a stochastic way to form a pilot vector. The target result can receive the characteristics of the giver solution or mutant through this crossover process. Uniform crossover and exponential crossover are two often utilized crossover operators. A crossover rate (CR) with a value between [0,1] controls the uniform crossover method. The uniform crossover trial solution can be described as follows in (Equation (8.11)):

$$Z_i^t = \begin{cases} Y_{i,j}^t & \text{if } rand_{i,j}[0,1] \le CR \text{ or } j = k \\ X_{i,j}^t & \text{Otherwise} \end{cases} \tag{8.11}$$

where $rand_{i,j}$ is a random value that falls between 0 and 1 and $k \in \{1,2..,D\}$ is a measurement guide chosen at random to confirm that as a minimum one-dimensional element of the pilot solution Z_i^t is a descendant of the $Y_{i,j}^t$ giver vector.

For exponential crossover, the basic concept of the dimensional indices for a reference vector to execute crossover with the mutant or donor vector is chosen at random from the integers $n \in \{1,2,........,D\}$. The quantity of dimension elements to be acquired from the donating or mutants vector to generate the pilot solution is denoted by the integer $L \in \{1,2,...,D\}$. The trial solution, Z_i^t, can be found from (Equation (8.12)) using the values of n and L as follows:

$$Z_i^t = \begin{cases} Y_{i,j}^t & \text{if } j = \langle n \rangle_D, \ \langle n+1 \rangle_D, ..., \langle n+L-1 \rangle_D \\ X_{i,j}^t & \text{Otherwise} \end{cases} \tag{8.12}$$

where the D modulus function is denoted by $\langle . \rangle_D$. Various types of optimization algorithms, particularly ones having links amongst neighboring choice variables, are said to work better with logarithmic crossover.

8.2.6.2.4 Selection

This procedure allows DE to identify whether an objective (parent) or a pilot (offspring) solution will survive in the following repetition (X_i^{t+1}) of the examination process though keeping DE's populace extent constant. Once the following generation's population has been established, the cyclic progressions of mutation, crossover, and selection are repeated till the termination requirements are met. There are two sorts of selection: local and worldwide (Xiao et al. 2019). The mathematical description of DE's selection process is as follows:

$$X_i^{t+1} = \begin{cases} Z_i^t & \text{if } f(Z)_i^t \leq fX_i^t \\ X_i^t & \text{Otherwise} \end{cases} \tag{8.13}$$

where *f(.)* is an operator for determining an individual solution's fitness function or predicted value. In the next iteration, if the newest trial vector of Z_i^t delivers a well impartial function value, the existing objective vector X_i^t would be substituted by Z_i^t. The DE's selection process can be done in both synchronous and asynchronous ways. In synchronous mode, the DE population can be updated all at once, whereas in asynchronous mode, the DE population can be updated one at a time.

In brief differential evolution experiment is based on the population used for global optimization on a constant sphere, categorized by easiness, efficiency, and strength (Bhattacharya et al. 2011). It works on the principle to construct a transmuted vector for every element at every generation. This mutant vector was created by an alteration method by increasing the distances between two arbitrarily nominated components in the population. Based on the transmuted vector, a pilot vector was created by the crossover process to associate the elements from the present element and the transmuted vector. The pilot vector fight with the related component of the present generation and the top element based on the objective function is shifted to the next population. A summary of a modified DE algorithm (Bhattacharya et al. 2011) is illustrated as follows:

Step 1: Prepare the generation;

Step 2: Estimate the generation;

Step 3: Assemble a new population in which all single aspirant is produced in equivalent conferring:

  i.   Arbitrary choose three different aspirants, r1; r2; r3 from the generation which is diverse from i;

  ii. Prepare a pilot vector;

  iii. Incorporate the matches of this vector by possibility control parameter, using single minimum match;

  iv. If the nominee solution is unacceptable, change its unacceptable matches by changing the nominee to the adjacent one;

  v.   Estimate the nominee solution;

  vi. If the nominee solution is good as the existing individual, then apply it to the next population;

Step 4:   Loop step 3 till the cessation condition is found.

Computational-based differential evolution is an effective method that can be employed to optimize the fermentation medium for biologically important molecules' production. To run the differential evolution in MATLAB, three codes are required viz. DE, Sphere, and main. The code obtained from Mostapha Kalami Heris, DE in MATLAB (URL: https://yarpiz.com/231/ypea107-differential-evolution), (Yarpiz, 2015) is as follows:

DE

```
%
% Copyright (c) 2015, Mostapha Kalami Heris & Yarpiz (www.yarpiz.
com)
% All rights reserved. Please read the "LICENSE" file for license terms.
%
% Project Code: YPEA107
% Project Title: Implementation of Differential Evolution (DE) in
MATLAB
% Publisher: Yarpiz (www.yarpiz.com)
%
% Developer: Mostapha Kalami Heris (Member of Yarpiz Team)
%
% Cite as:
% Mostapha Kalami Heris, Differential Evolution (DE) in MATLAB
(URL: https://yarpiz.com/231/ypea107-differential-evolution),
Yarpiz, 2015.
%
% Contact Info: sm.kalami@gmail.com, info@yarpiz.com
%

clc;
clear;
close all;

%% Problem Definition

CostFunction = @(x) Sphere(x); % Cost Function

nVar = 20;    % Number of Decision Variables

VarSize = [1 nVar]; % Decision Variables Matrix Size
```

```
VarMin = −5;   % Lower Bound of Decision Variables
VarMax = 5;    % Upper Bound of Decision Variables

%% DE Parameters

MaxIt = 1000; % Maximum Number of Iterations

nPop = 50; % Population Size

beta_min = 0.2; % Lower Bound of Scaling Factor
beta_max = 0.8; % Upper Bound of Scaling Factor

pCR = 0.2; % Crossover Probability

%% Initialization
empty_individual.Position = [];
empty_individual.Cost = [];
BestSol.Cost = inf;

pop = repmat(empty_individual, nPop, 1);

for i = 1:nPop

    pop(i).Position = unifrnd(VarMin, VarMax, VarSize);
    pop(i).Cost = CostFunction(pop(i).Position);
    if pop(i).Cost<BestSol.Cost
        BestSol = pop(i);
    end

end

BestCost = zeros(MaxIt, 1);

%% DE Main Loop

for it = 1:MaxIt
 for i = 1:nPop

   x = pop(i).Position;

   A = randperm(nPop);
```

```
A(A == i) = [];

    a = A(1);
    b = A(2);
    c = A(3);

    % Mutation
    %beta = unifrnd(beta_min, beta_max);
    beta = unifrnd(beta_min, beta_max, VarSize);
    y = pop(a).Position+beta.*(pop(b).Position-pop(c).Position);
    y = max(y, VarMin);
                y = min(y, VarMax);

    % Crossover
    z = zeros(size(x));
    j0 = randi([1 numel(x)]);
    for j = 1:numel(x)
        if j == j0 || rand <= pCR
            z(j) = y(j);
        else
            z(j) = x(j);
        end
    end

    NewSol.Position = z;
    NewSol.Cost = CostFunction(NewSol.Position);

    if NewSol.Cost<pop(i).Cost
        pop(i) = NewSol;

    if pop(i).Cost<BestSol.Cost
            BestSol = pop(i);
        end
    end

    end
% Update Best Cost
    BestCost(it) = BestSol.Cost;
```

```
    % Show Iteration Information
    disp(['Iteration ' num2str(it) ': Best Cost = ' num2str(BestCost(it))]);

end

%% Show Results

figure;
%plot(BestCost);
semilogy(BestCost, 'LineWidth', 2);
xlabel('Iteration');
ylabel('Best Cost');
grid on;
```

Sphere

```
%
% Copyright (c) 2015, Mostapha Kalami Heris & Yarpiz (www.yarpiz.
com)
% All rights reserved. Please read the "LICENSE" file for license
terms.
%
% Project Code: YPEA107
% Project Title: Implementation of Differential Evolution (DE) in
MATLAB
% Publisher: Yarpiz (www.yarpiz.com)
%
% Developer: Mostapha Kalami Heris (Member of Yarpiz Team)
%
% Cite as:
% Mostapha Kalami Heris, Differential Evolution (DE) in MATLAB
(URL: https://yarpiz.com/231/ypea107-differential-evolution),
Yarpiz, 2015.
%
% Contact Info: sm.kalami@gmail.com, info@yarpiz.com
%

function z = Sphere(x)

    z = sum(x.^2);

end
```

Main

```
%
% Copyright (c) 2015, Mostapha Kalami Heris & Yarpiz (www.yarpiz.
com)
% All rights reserved. Please read the "LICENSE" file for license
terms.
%
% Project Code: YPEA107
% Project Title: Implementation of Differential Evolution (DE) in
MATLAB
% Publisher: Yarpiz (www.yarpiz.com)
%
% Developer: Mostapha Kalami Heris (Member of Yarpiz Team)
%
% Cite as:
% Mostapha Kalami Heris, Differential Evolution (DE) in MATLAB
(URL: https://yarpiz.com/231/ypea107-differential-evolution),
Yarpiz, 2015.
%
% Contact Info: sm.kalami@gmail.com, info@yarpiz.com
%

de;
```

8.2.7 Validation of the Experiment

The validation of the model obtained by CCD was done by executing the experiment under the predicted conditions. Samples were collected every 24 h and analyzed for chitinase activity. The outcomes of the validated experiment were compared with results attained by using basal media.

8.3 Results and Discussion

8.3.1 Plackett–Burman Design

The PBD was used to screen the important variables that can affect chitinase activity. Nineteen variables with 20 experiments were accompanied and it was discovered that there was a major difference in the level of chitinase production; chitinase activity was found to be lowest and maximum in the range of 7.51–34.18 U/L (Table 8.2). This variation in activity evidenced the importance of the optimization process for medium constituents to attain the higher activity of chitinase produced by *T. lanuginosus* by submerged fermentation (Zeng et al. 2018). The results obtained from PBD experiments were evaluated by employing an ANOVA (Table 8.4) to find out the significant variables. In 15 variables, the most significant variables found for chitinase production are yeast extract, Tween 80, glucose, and colloidal chitin. The high F-value (49.21) infers the significance of the model. This resembles the model P-value ($>F$) is less than 0.0001. It means there is only 0.01% chance that the large F-value (49.21) of the model occurred due to the noise. The R^2 value of 0.9883 revealed that the model could describe 98.83% variation in data. Moreover, a very less coefficient of variance, i.e., 8.05%, also implies that the model is reliable. All the variables with P-values less than 0.05 were concluded as significant for chitinase production. By using the Pareto chart, the negatively and positively affecting variables were selected (Figure 8.2). It was observed that colloidal chitin, $CaCl_2$, Tween 80, yeast extract, KCl, and $MnSO_4$ was showing a positive effect, and $FeSO_4$, NaH_2PO_4, and $CoCl_2$ show a negative effect on chitinase production, whereas glucose, Tween 80, yeast extract, and KH_2PO_4 show a positive effect and $CaCl_2$, $MnSO_4$, NaH_2PO_4,

TABLE 8.4

Experimental Effects of the Plackett–Burman Design for Chitinase Production by *T. lanuginosus* Were Analyzed Using the Analysis of Variance

Source	Sum of Squares	df	Mean Square	F-Value	P-Value	
Model	1317.92	12	109.83	49.21	< 0.0001	Significant
A-Colloidal chitin	437.63	1	437.63	196.09	< 0.0001	
D-Yeast extract	49.25	1	49.25	22.07	0.0022	
F-Na_2HPO_4	117.59	1	117.59	52.69	0.0002	
H-$MnSO_4$	16.31	1	16.31	7.31	0.0305	
J-$CaCl_2$	81.26	1	81.26	36.41	0.0005	
K-KCl	17.41	1	17.41	7.80	0.0268	
M-$FeSO_4$	222.21	1	222.21	99.57	< 0.0001	
N-$CoCl_2$	31.31	1	31.31	14.03	0.0072	
O-Tween 80	192.77	1	192.77	86.38	< 0.0001	
Q-Dummy 1	19.44	1	19.44	8.71	0.0214	
S-Dummy 3	68.89	1	68.89	30.87	0.0009	
T-Dummy 4	63.87	1	63.87	28.62	0.0011	
Residual	15.62	7	2.23			
Cor Total	1333.55	19				

Note: Significant model terms are indicated by values of "Prob $> F < 0.0500$".

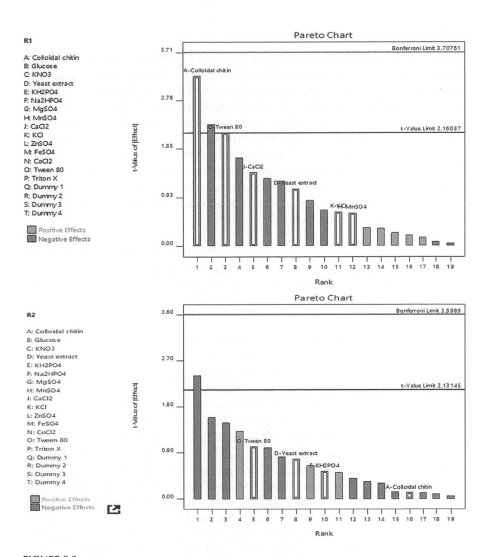

FIGURE 8.2
Pareto chart showing the negatively and positively affecting variables for chitinase activity (R1) and biomass (R2) obtained by Placket–Burman design.

$CoCl_2$, $FeSO_4$, $MgSO_4$, and $ZnSO_4$ show a negative effect on biomass. By substituting the variables showing a negative effect on the chitinase activity and biomass, four variables were selected to employ the central composite design for further optimization. In general, and to screen the significant medium components to formulate the optimum media for chitinase production, the PBD has been successfully used by several researchers (Mishra et al. 2012; Patel et al. 2007; Nawani and Kapadnis 2005).

8.3.2 Central Composite Design

An economic production medium is required for the cost-effective production process of biocatalysts used in industries. Optimization studies for enhanced production of enzymes are improved by employing an RSM that involves selecting a significant variable and optimal concentration for the best outcomes. The four variables ($k = 4$) *viz.* yeast extract, Tween 80, glucose, and colloidal chitin selected by PBD were analyzed for their effect and interaction at five coded levels with the help of a central composite design (Table 8.3). With the central composite design results, ANOVA was performed, and a second-order polynomial equation (Suryawanshi and Jujjawarapu 2020) was constructed to explain the output of chitinase as a function of the analyzed variables (Equation (8.2)).

$$\begin{aligned} R = {} & 73.75 + 2.21 \times A - 0.79 \times B + 3.58 \times C + 3.66 \times D + 1.43 \times A \times B \\ & + 0.52 \times A \times C - 2.06 \times A \times D - 0.54 \times B \times C - 1.42 \times B \times D \\ & + 2.14 \times C \times D + 2.56 \times A^2 + 2.92 \times B^2 + 4.2 \times C^2 \\ & + 2.17 \times D^2 \end{aligned} \quad (8.14)$$

R stands for chitinase activity (U/L), and A, B, C, and D stand for colloidal chitin, glucose, Tween 80, and yeast extract, respectively.

ANOVA was employed for the chitinase activity, where the large F-value of 2.71% and the P-value of the model is 0.0334 was showing that the model is fit, and there are only 3.24% chances that the large F-value (2.71%) of the model was occurred due to the noise (Table 8.5). The R^2 value of 0.7170 revealed that the model could describe a 71.70% variation in data. Moreover, a very less coefficient of variance, i.e., 8.12%, implies that the model is sufficient to optimize the medium components. Adequate precision is a quantity of response to the deviation (signal to noise) ratio, and its value larger than 4 is usually appropriate. In this study, the adequate precision value of 5.1891 has shown a satisfactory indication that the model can be employed to direct the design space. Model terms that have a P-value < 0.05 were deliberated as significant. The model terms that have a P-value > 0.1 are deliberated as insignificant. ANOVA results (Table 8.5) showed that the Tween 80, yeast extract, and the squared terms of glucose and Tween 80 significantly affected the chitinase activity. Among them, Tween 80 was found the most affecting variable for chitinase activity. The contour plot showed the effect of yeast extract and Tween 80 on chitinase activity (Figure 8.3).

However, the effect of the interaction of all variables was insignificant. Statistical optimization for medium composition to enhance the chitinase activity was effectively used by various researchers (Jha and Modi 2018; Rawway et al. 2018; Kumar et al. 2018; Sukalkar et al. 2018, Wang et al. 2014; De-hui et al. 2011). Exploiting the RSM instead of the OFAT process gives the advantage of analyzing the influence due to interaction among the

TABLE 8.5

ANOVA for the Quadratic Response Surface for Chitinase Production by *T. lanuginosus*

Source	Sum of Squares	df	Mean Square	F-Value	P-Value	
Model	1734.78	14	123.91	2.71	0.0324	Significant
A-Colloidal chitin	117.49	1	117.49	2.57	0.1295	
B-Glucose	14.91	1	14.91	0.3265	0.5762	
C-Tween 80	308.44	1	308.44	6.76	0.0201	
D-Yeast extract	322.17	1	322.17	7.06	0.0180	
AB	32.65	1	32.65	0.7150	0.4111	
AC	4.26	1	4.26	0.0932	0.7643	
AD	68.09	1	68.09	1.49	0.2409	
BC	4.70	1	4.70	0.1029	0.7528	
BD	32.11	1	32.11	0.7032	0.4149	
CD	73.24	1	73.24	1.60	0.2246	
A^2	180.14	1	180.14	3.95	0.0656	
B^2	234.62	1	234.62	5.14	0.0386	
C^2	493.03	1	493.03	10.80	0.0050	
D^2	128.97	1	128.97	2.82	0.1135	
Residual	684.88	15	45.66			
Lack of Fit	684.65	10	68.47	1521.88	< 0.0001	Significant
Pure Error	0.2249	5	0.0450			
Cor Total	2419.66	29				

Note: Significant model terms are indicated by values of "Prob > F < 0.0500".

independent factors on the response. There is no effect found on the interaction of independent factors. The predicted values (%) obtained by CCD were yeast extract, 6; Tween 80, 0.07; glucose, 0.25, and colloidal chitin 1.4. The predicted highest chitinase activity of 94.38 U/L. Validation was performed with these conditions to find out the significance of the model. The regression equation was further applied in DE to estimate the variables' optimum values with enhanced activity of chitinase and compare the model results with differential evolution.

8.3.3 Differential Evolution (DE)

The polynomial equation obtained by CCD-RSM was applied in the algorithm of differential evolution. The three significant regulatory parameters of differential evolution, *viz.* scaling factor, crossover operator (CR), and population (NP) were chosen by an efficient analysis. In this study, the magnitude of the population was 50. For the selection of CR, the differential evolution algorithm was performed with different CR values from 0.0 to 0.6. The other

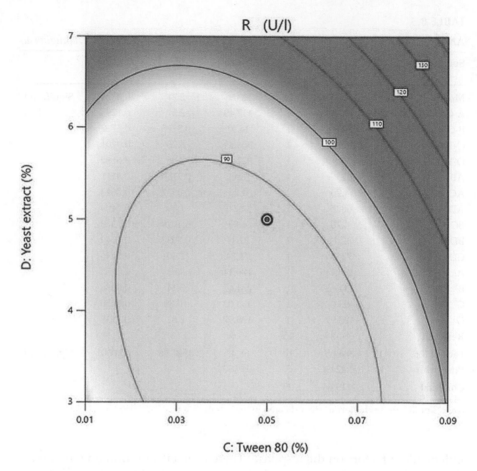

FIGURE 8.3
Contour plot depicting the effect of yeast extract and Tween 80 on chitinase activity.

parameters were kept as follows: beta max, 0.8; beta min, 0.2; C, 25; max iteration, 50; the number of populations, 50; with four variables. With all the different CR values, chitinase's maximum activity was obtained as 122.84 U/L with the optimum value of variables (%), i.e., yeast extract, 4; Tween 80, 0.03; glucose, 0.1790, and colloidal chitin, 1 (Figure 8.4).

8.3.4 Experiment Model Validation

The experimental model obtained by CCD RSM was validated and compared with the basal media. The highest activity of chitinase under predicted experimental conditions was obtained at 92.496 U/L, whereas the highest activity obtained by the basal media was 42.75 U/L (Figure 8.5). It shows that the activity of chitinase was improved just two-fold by the optimized media.

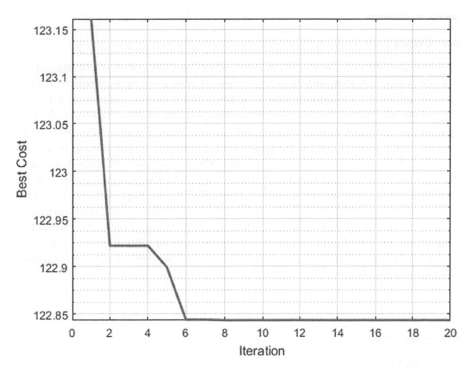

FIGURE 8.4
Objective function value variation with the 20 iterations; the number of populations is 50, and CR is 0.6.

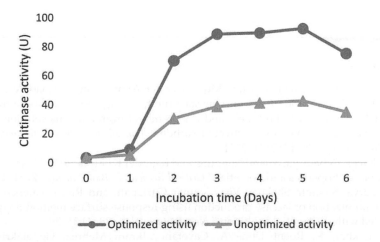

FIGURE 8.5
Comparison of improved chitinase activity with the activity obtained by basal media.

8.4 Conclusion

The components of the medium were optimized to create a chitinase production medium. Plackett–Burman's design was used to determine significant variables. CCD-RSM successfully finds the modeling and interaction of the variables. The second-order polynomial equation obtained by RSM was applied in differential evolution. RSM results have shown the model is fit, and the validation of the experiment confirmed it. With the predicted model by RSM, the chitinase activity was improved two-fold compared to basal media. The activity obtained was much closer to the activity predicted by the RSM model. By the optimization, through DE, the highest activity of chitinase was found 122.84 U/L with the following concentration of variables: yeast extract 4%, Tween 80 0.03%, glucose 0.1790%, and colloidal chitin 1%, which was much higher as compared to activity obtained by CCD-RSM. It proves that computational differential evolution can be an effective optimization technique.

Acknowledgments

Nisha Suryawanshi would like to express her gratitude to the CSIR New Delhi for the fellowship and the National Institute of Technology, Raipur, India, for assisting her with her study.

References

Alslaibi, Tamer M, Ismail Abustan, Mohd Azmier Ahmad, and Ahmad Abu Foul. 2013. "Application of response surface methodology (RSM) for optimization of Cu^{2+}, Cd^{2+}, Ni^{2+}, Pb^{2+}, Fe^{2+}, and Zn^{2+} removal from aqueous solution using microwaved olive stone activated carbon." *Journal of Chemical Technology & Biotechnology* 88 (12):2141–2151.

Bhattacharya, Debaditya, Anand Nagpure, and Rajinder K Gupta. 2007. "Bacterial chitinases: Properties and potential." *Critical Reviews in Biotechnology* 27 (1):21–28.

Bhattacharya, Sukanta Shekhar, Vijay Kumar Garlapati, and Rintu Banerjee. 2011. "Optimization of laccase production using response surface methodology coupled with differential evolution." *New Biotechnology* 28 (1):31–39.

Chandrasekhar, K, Roent Dune A. Cayetano, Ikram Mehrez, Gopalakrishnan Kumar, and Sang-Hyoun Kim. 2020. "Evaluation of the biochemical methane potential of different sorts of Algerian date biomass." *Environmental Technology and Innovation* 20:101180.

Chandrasekhar, K., A. Naresh Kumar, Gopalakrishnan Kumar, Dong-Hoon Kim, Young-Chae Song, and Sang-Hyoun Kim. 2021. "Electro-fermentation for biofuels and biochemicals production: Current status and future directions." *Bioresource Technology* 323:124598.

Chen, Junpeng, Yangdongfang An, Ashok Kumar, and Ziduo Liu. 2017. "Improvement of chitinase Pachi with nematicidal activities by random mutagenesis." *International Journal of Biological Macromolecules* 96:171–176.

Dahiya, Neetu, Rupinder Tewari, Ram Prakash Tiwari, and Gurinder Singh Hoondal. 2005. "Chitinase production in solid-state fermentation by Enterobacter sp. NRG4 using statistical experimental design." *Current Microbiology* 51 (4):222–228.

Das, Swagatam, and Ponnuthurai Nagaratnam Suganthan. 2010. "Differential evolution: A survey of the state-of-the-art." *IEEE Transactions on Evolutionary Computation* 15 (1):4–31.

De-Hui, Dai, Li Wei, Hu Wei-Lian, and Sa Xiao-Ying. 2011. "Effect of medium composition on the synthesis of chitinase and chitin deacetylase from thermophilic Paenibacillus sp. Hul." *Procedia Environmental Sciences* 8:620–628.

Dhillon, Gurpreet Singh, Satinder Kaur Brar, Surinder Kaur, Jose R Valero, and Mausam Verma. 2011. "Chitinolytic and chitosanolytic activities from crude cellulase extract produced by A. niger grown on apple pomace through koji fermentation." *Journal of Microbiology and Biotechnology* 21 (12):1312–1321.

Doddapaneni, Kiran Kumar, Radhika Tatineni, Ravichandra Potumarthi, and Lakshmi Narasu Mangamoori. 2007. "Optimization of media constituents through response surface methodology for improved production of alkaline proteases by Serratia rubidaea." *Journal of Chemical Technology & Biotechnology: International Research in Process, Environmental & Clean Technology* 82 (8):721–729.

Eswari, Jujjavarapu Satya, Mohan Anand, and Chimmiri Venkateswarlu. 2013. "Optimum culture medium composition for rhamnolipid production by pseudomonas aeruginosa AT10 using a novel multi-objective optimization method." *Journal of Chemical Technology & Biotechnology* 88 (2):271–279.

Jha, Sneha, and H.A. Modi. 2018. "Statistical optimization of chitinase production by Streptomyces rubiginosus SP24 and efficacy of purified chitinase to control root-knot nematode infection in Vigna radiata under controlled conditions." *Chemical and Biological Technologies in Agriculture* 5 (1):20.

Karn, Santosh Kr, and Jizhou Duan. 2017. *Perspective on Microbial Influenced Corrosion.*

Karn, Santosh Kumar, and Awanish Kumar. 2019. "Protease, lipase and amylase extraction and optimization from activated sludge of pulp and paper industry."

Kenneth, Storn R., and Rainer M. Storn. 1997. "Differential evolution-a simple and efficient heuristic for global optimization over continuous spaces." *Journal of Global Optimization* 11 (4):341–359.

Khandelwal, Amit Kumar, Vinod Kumar Nigam, Bijan Choudhury, Medicherla Krishna Mohan, and Purnendu Ghosh. 2007. "Optimization of nitrilase production from a new thermophilic isolate." *Journal of Chemical Technology & Biotechnology: International Research in Process, Environmental & Clean Technology* 82 (7):646–651.

Kumar, D. Praveen, Rajesh Kumar Singh, P.D. Anupama, Manoj Kumar Solanki, Sudheer Kumar, Alok K. Srivastava, Pradeep K. Singhal, and Dilip K. Arora. 2012. "Studies on exo-chitinase production from Trichoderma asperellum UTP-16 and its characterization." *Indian Journal of Microbiology* 52 (3):388–395.

Kumar, Manish, Amandeep Brar, V. Vivekanand, and Nidhi Pareek. 2017. "Production of chitinase from thermophilic Humicola grisea and its application in production of bioactive chitooligosaccharides." *International Journal of Biological Macromolecules* 104:1641–1647.

Kumar, Manish, Amandeep Brar, V. Vivekanand, and Nidhi Pareek. 2018. "Process optimization, purification and characterization of a novel acidic, thermostable chitinase from Humicola grisea." *International Journal of Biological Macromolecules* 116:931–938.

Liu, Chao-Lin, Tsung-Han Lin, and Ruey-Shin Juang. 2013. "Optimization of recombinant hexaoligochitin-producing chitinase production with response surface methodology." *International Journal of Biological Macromolecules* 62:518–522.

Mak, Kenneth W.Y., Miranda G.S. Yap, and Wah Koon Teo. 1995. "Formulation and optimization of two culture media for the production of tumour necrosis factor-β in Escherichia coli." *Journal of Chemical Technology & Biotechnology: International Research in Process, Environmental AND Clean Technology* 62 (3):289–294.

Mishra, P., P.R. Kshirsagar, S.S. Nilegaonkar, and S.K. Singh. 2012. "Statistical optimization of medium components for production of extracellular chitinase by Basidiobolus ranarum: A novel biocontrol agent against plant pathogenic fungi." *Journal of Basic Microbiology* 52 (5):539–548.

Montgomery, D.C. 2001. *Design and Analysis of Experiments*, 5th edn, John Wiley & Sons, Inc.: New York.

Montgomery, D.C. 2003. *Design and Analysis of Experiments*. John Wiley & Sons (ASIA) Pvt. Ltd.

Nawani, N.N., and B.P. Kapadnis. 2005. "Optimization of chitinase production using statistics based experimental designs." *Process Biochemistry* 40 (2):651–660.

Neri, Ferrante, and V. Tirronen 2010. "Recent advances in differential evolution: A survey and experimental analysis." *Artificial Intelligence Review* 33 (1–2):61–106.

Oranusi, Nathaniel A., and A.P. Trinci. 1985. "Growth of bacteria on chitin, fungal cell walls and fungal biomass, and the effect of extracellular enzymes produced by these cultures on the antifungal activity of amphotericin B." *Microbios* 43 (172):17–30.

Pant, Millie, Hira Zaheer, Laura Garcia-Hernandez, and Ajith Abraham. 2020. "Differential evolution: A review of more than two decades of research." *Engineering Applications of Artificial Intelligence* 90:103479.

Patel, Bharat, Vipul Gohel, and Bhairavsinh Raol. 2007. "Statistical optimisation of medium components for chitinase production byPaenibacillus sabina strain JD2." *Annals of Microbiology* 57 (4):589–597.

Plackett, Robin L., and J. Peter Burman. 1946. "The design of optimum multifactorial experiments." *Biometrika* 33 (4):305–325.

Price, Kenneth, Rainer M. Storn, and Jouni A. Lampinen. 2006. *Differential Evolution: A Practical Approach to Global Optimization*. Springer Science & Business Media.

Purushotham, Pallinti, P.V.S.R.N. Sarma, and Appa Rao Podile. 2012. "Multiple chitinases of an endophytic Serratia proteamaculans 568 generate chitin oligomers." *Bioresource Technology* 112:261–269.

Qin, A., V. Huang, and P. Suganthan. 2009. "Differential evolution algorithm with strategy adaptation for global numerical optimization." *IEEE Transactions on Evolutionary Computation* 13:398–417.

Rawway, Mohammed, Ehab Aly Beltagy, Usama Mohamed Abdul-Raouf, Mohamed Ahmed Elshenawy, and Mahmud Saber Kelany. 2018. "Optimization of process parameters for chitinase production by a marine Aspergillus flavus MK20." *Journal of Ecology of Health & Environment* 6:1–8.

Roberts, Walden K., and Claude P. Selitrennikoff. 1988. "Plant and bacterial chitinases differ in antifungal activity." *Microbiology* 134 (1):169–176.

Roudi, Anita Maslahati, Hesam Kamyab, Shreeshivadasan Chelliapan, Veeramuthu Ashokkumar, Ashok Kumar, Krishna Kumar Yadav, and Neha Gupta. 2020. "Application of response surface method for total organic carbon reduction in leachate treatment using Fenton process." *Environmental Technology and Innovation* 19:101009.

Sahai, A.S., and M.S. Manocha. 1993. "Chitinases of fungi and plants: Their involvement in morphogenesis and host-parasite interaction." *FEMS Microbiology Reviews* 11 (4):317–338.

Singh, Suren, Andreas M Madlala, and Bernard A. Prior. 2003. "Thermomyces lanuginosus: Properties of strains and their hemicellulases." *FEMS Microbiology Reviews* 27 (1):3–16.

Storn, Rainer, and Kenneth Price. 1997. "Differential evolution–A simple and efficient heuristic for global optimization over continuous spaces." *Journal of Global Optimization* 11 (4):341–359.

Sukalkar, Swati R., Tukaram A. Kadam, and Hemlata J. Bhosale 2018. "Optimization of chitinase production from streptomyces macrosporeus M1."

Suryawanshi, Nisha, and Satya Eswari Jujjawarapu. 2020. "Chitin from seafood waste: Particle swarm optimization and Neural network study for the improved chitinase production." *Journal of Chemical Technology and Biotechnology.* doi: https://doi.org/10.1002/jctb.6656.

Vaidya, R.J., S.L.A. Macmil, P.R. Vyas, and H.S. Chhatpar. 2003. "The novel method for isolating chitinolytic bacteria and its application in screening for hyper-chitinase producing mutant of Alcaligenes xylosoxydans." *Letters in Applied Microbiology* 36 (3):129–134.

Wang, Kai, Pei-sheng Yan, and Li-xin Cao. 2014. "Chitinase from a novel strain of Serratia marcescens JPP1 for biocontrol of aflatoxin: Molecular characterization and production optimization using response surface methodology." *BioMed Research International* 2014.

Wasli, Azaliza Safarida, Madihah Md Salleh, Suraini Abd-Aziz, Osman Hassan, and Nor Muhammad Mahadi. 2009. "Medium optimization for chitinase production from Trichoderma virens using central composite design." *Biotechnology and Bioprocess Engineering* 14 (6):781–787.

Xiao, Peng, Dexuan Zou, Zhenglong Xia, and Xin Shen. 2019. "Multi-strategy different dimensional mutation differential evolution algorithm." *AIP Conference Proceedings.*

Yu, X., S.G. Hallett, J. Sheppard, and A.K. Watson. 1997. "Application of the Plackett-Burman experimental design to evaluate nutritional requirements for the production of Colletotrichum coccodes spores." *Applied Microbiology and Biotechnology* 47 (3):301–305.

Zeng, Wei, Guiguang Chen, Yange Wu, Mengna Dong, Bin Zhang, and Zhiqun Liang. 2018. "Nonsterilized fermentative production of poly-γ-glutamic acid from cassava starch and corn steep powder by a thermophilic Bacillus subtilis." *Journal of Chemical Technology & Biotechnology* 93 (10):2917–2924.

Zhang, Meng. 2014. "The chitinolytic enzyme system of the compost-dwelling thermophilic fungus Thermomyces lanuginosus."

9

Artificial Bee Colony for Optimization of Process Parameters for Various Enzyme Productions

Dheeraj Shootha, Pooja Thathola, and Khashti Dasila
Kumaun University, Bhimtal, India

CONTENTS

9.1 Introduction

Optimization is the essential and crucial part of many optimization processes of designing, whether it is designed for business planning or machine learning. The optimization purpose could be anything; it is used to minimize the consumption of energy, time of the experiments, waste, costs, and other environmental impacts (pH, Temperature, agitation speed, etc.) or to maximize the response, i.e., performance, profit, efficiency, and sustainability of work. In real-world applications, there is limited time and money; so for this, there are many solutions to overcome this problem. One of them is experimental designing, which is subject to a wide range of complex constraints (Tsai et al. 2009). To solve these problems, optimization techniques are needed. As in reissue problems, most issues are in nonlinear terms for objective functions, so there is a need to use these essential sophisticated tools for the optimization that deals with such types of problems. Accumulation and the estimation purposes of this objective are further time-consuming processes – the major problem in designing fields. Therefore, among all the naturally inspired algorithms, Artificial Bee Colony Optimization (ABC) is the most common swarm intelligence (Karaboga and Basturk 2007) inspired by a field of biological computing that applies concepts from the combined behavior of swarms (i.e., bees) to solve problems in different areas like optimization of process parameters, etc. The researchers have been attracted to the combined quick performance of bees in these algorithms. The cumulative behavior of these bees is called swarm behavior. Entomologists study the engineer's and the biological swarm's behavior and apply these models to the framework toward complex problems. Swarm intelligence is artificial intelligence that mainly deals with the collective behavior of swarms by interactions with individuals without any supervision. Any effort to algorithm design before distributing this is used for the solution to the problem of social animals (Zhu and Kwong 2010). The swarm intelligence's main advantage is scalability, tolerance of the fault, speed, adaptation, and parallelism. Labor division and self-organization are the two critical parameters of swarm intelligence. Encountered units of local stimuli may respond individually in a self-organizing system and globally accomplish charge via labor separation without centralized supervision. The whole arrangement should have the ability to adapt to inner and outer changes capably. Four basic properties have been characterized on which self-organization mainly depends: positive feedback, negative feedback, multiple interactions, and fluctuations (Mohapatra et al. 2017). Positive feedback indicates that another individual by some instruction, like dancing bees, leads to another food source site. Negative feedback indicates exhaustion of the food source. Fluctuations refer to the random behaviors of each individual in the exploration of new sites (Beg et al. 2012). ABC may apply to various problems, including the training of artificial neural networks (ANN), for the designing of the infinite impulse response (IIR) (Zhang et al. 2010), for solving the problems related to constrained optimization (Sonmez 2011), and in the tertiary structure's prediction of proteins (Tsai et al. 2009). ABC algorithm performance toward the optimization of several other algorithms such as Particle Swarm Optimization (PSO), Genetic Algorithms (GA), Particle Swarm Inspired Evolutionary Algorithm (PS-EA), Differential Evolution (DE), and different evolutionary strategies (Liu et al. 2018). All the above-mentioned parameters have the ability to deal with specific kinds of clusters. However, there are many problems in this field that need to be solved (Chang and He 2014; Garoudja et al. 2015). First, deeper knowledge about the algorithms needs to understand the empirical sets. The performance of the algorithms mainly depends on the quality of the parameters significantly. Second, some codes can work fine with a limited kind of potential but not with others. Parameters are optimized for a high amount of enzyme production (Babaeizadeh and Ahmad 2014). Among all the different parameters for enzyme production, the condition at which the processes performed best should be optimized (Zhu and Kwong 2010). Usually, the process of higher enzyme production may be optimized using different variables, i.e., temperature, reaction time, pH, and nutrient concentration (Ilie and Bădică 2013). To achieve higher enzyme production, the optimization of the variables is generally not possible for "one variable at a time" approach (Baykaso et al. 2007). In biotechnological processes, researchers started statistical optimization and nature-based optimization approaches to increase the production of enzymes (Agrawal et al. 2016; Garlapati and Banerjee 2010; Liu and Tang 2018; Mahapatra et al. 2009). To support the facts mentioned earlier, research has to start utilizing the different natural development-based easy computing approaches to optimize enhanced enzyme production (El-Abd 2011). ABC is one of the optimization techniques that involves the bees for optimization purposes. ABC is applied to the broad area for optimization purposes. There are many reports published on ABC optimization from 2005 to 2022, as shown in Figure 9.1.

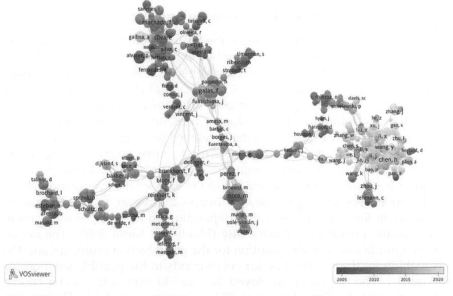

FIGURE 9.1
Publications related to ABC optimization from 2005 to 2021.

9.2 ABC Algorithm Motivations

ABC algorithms are commonly known swarm methods for optimization which is mainly inspired by bees foraging behavior (Karaboga et al. 2014). Throughout the process of optimization, the ABC algorithm is featured by local and global exploration (Li et al. 2015). However, the local accuracy search for the ABC is not very acceptable (Zhu and Kwong 2010), leading to a large number of remedies proposed for this imperfection. The most proposed ABC variants have been deemed for adjustment of the local utilization of the equations for the achievement of a good intensity search. For this, three major pathways were taken: (i) adoption of novel principle from the external world, (ii) integration of the ABC with the other metaheuristics, and (iii) adjustment of the self-adaptivity for the local search concentration (Gao et al. 2014). It is tough to analyze the outperformance of ABC due to its statistical significance in comparative numerical experiments (Xiang and An 2013). In-depth mathematical analysis in the research made these algorithms precious. The difference with the first two variation options – and that ABC variants are self-adaptive (i.e., third method) – are understandable intuitively (because of no involvement of outside-world, complicated principles), and continue the novel framework in the conventional algorithm.

9.3 Optimization of Artificial Bee Colony Algorithm

Honey bees can effectively determine the highest quality of food sources in nature. Hence, the intelligent foraging bee behavior is finding a good solution for solving optimization-related problems (Li et al. 2015; Sathesh Kumar and Hemalatha 2014; Sharma et al. 2021; Zhang et al. 2021). In general, for searching the food sources, the honey bees' colonies are mainly separated into three kinds, i.e., employed bees, onlooker bees, and scout bees (Figure 9.2). For the nectar source exploitation, the employed bees are responsible (Bansal et al. 2011). They go to the beforehand food source position and give the information to onlooker bees about the quality of the food source. Then, onlooker bees wait in the hive and decide to explore the information about food sources. For finding the new source of the nectar, scout bees search randomly in the environment either depending on the internal motivation, based on the possible clues externally (Hakli and Kiran 2020). The nectar position implies the possible solution for the optimization problems, and the profitability of the source of nectar corresponds to the possible solution for the quality. The respective employed bee should exploit the nectar source (Hu et al. 2021; Kasihmuddin et al. 2021; Gautam et al. 2019). The sources

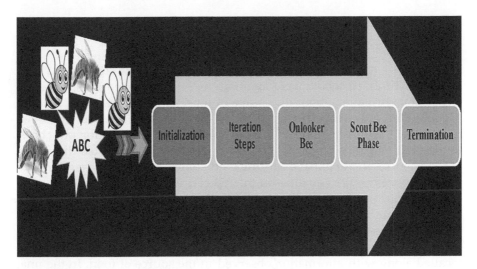

FIGURE 9.2
Honey bee colony according to the food source.

of nectar are equivalent to the number of employed bees and onlooker bees (Bansal et al. 2011). In this method, good solutions were obtained with employed bees and the spectator bees for the convergence improvement in the speed, and scout bees mainly improve the capacity to eliminate a local optimum (Babaoglu 2015).

9.4 Honey Bees' Foraging Behavior

Honey bee colonies can prey on a large number of food sources in a large field area and fly high up to 11 km to prey on the source of food. The employees in the colony are about an area of its members as collectors. Foraging processes is initiated by the search for a promising patch of the flower using scout bees (Brezočnik et al. 2018; Zhang et al. 2012). During harvest, the season colony maintains the percentage of scout bees. When bees identified the patch of the flower, they started looking for more with the hope of finding a better one. For the identification of better patches, the scout bees search randomly (Li et al. 2019; Harrison et al. 2018; Leu and Yeh 2012). Scout bees update their conspecifics resting in the hives for the quantification of the food source, which also includes the sugar content. Nectar was deposited by scout bees get on the "dance floor" in front of the hives and communication with the other bees was done by performing a type of dance known as "waggle dance" (Babaeizadeh and Ahmad 2014; Bansal et al. 2011; Chuang et al. 2008).

9.5 Iteration Steps in ABC for Optimization

The study of ABC follows mainly three most important steps (Akay and Karaboga 2012): (i) food source identification by employed bees; (ii) gathering the information about the source of food and choosing the quality of nectar – food source insurance is decided by onlooker bees, and employed bees gather the information and decide the eminence of nectar for scout bees, finding and making use of the ones in which they are interested; (iii) reach of food source. In the initial stage, the food source locality was selected randomly by bees and the qualities of their nectar were measured. Then, information about nectar sources is transferred by employed bees to onlooker bees which are waiting within beehives in the dance vicinity. After information distribution, every employed bee proceeds to the food source to check some stages in the previous cycle, when the food source position has been recalled along with information observed in the source of food. In the final stage, at the area of food source, the information from employed bees is retrieved by onlooker bees. Therefore, the information on food source quality in nectar is transferred through the employed bees. Then, the subsequent choice of other food sources mainly depends on the experiential information. Scout bees erratically generate a new source of food from deserted swap by onlooker bees.

9.5.1 Swarm Initialization

ABC algorithms have three major process parameters, i.e., the population which is generally a number of food sources, for the testing of subsequent food source numbers, they are mainly selected as deserted (limit), and criteria of termination, which mainly depend on the cycle numbers. Originally, Karaboga et al. (2014) demonstrated that the ABC number for a food source is equival to the number of employed and onlooker bees. In the beginning, it assumes a regularly distributed swarm source of food (SN), wherever each source of food x_i (where $i = 1, 2, ..., SN$) is mainly a D-dimensional vector. A subsequent equation generally used for every source of food is shown in Equation (9.1):

$$x_{i,j} = x_j^{\min} + \varphi_{i,j} \left(x_j^{\max} + x_j^{\min} \right) \tag{9.1}$$

where $\varphi_{i,j}$ is random (0,1) real numbers for equal distribution and $i = 1, 2, ..., SN$, $j = 1, ..., D$, and $x_j^{\max}$ and $x_j^{\min}$ are the dimensions for upper and lower bounds.

9.5.2 Onlooker Bee

Food source numbers designed for onlooker bees are like a food number for sourced employed bees. During this duration phase, all employed bees share new sources of food information about the availability with onlooker bees. Onlooker bees generally determine the probability of selection for every food source that is engendered by an employed bee (Gao et al. 2014). There are several schemes for calculating probability and they must include suitability. For each food source, the probability is usually based on its capability as shown in Equation (9.2):

$$P_i- = \frac{\text{fit}_i}{\sum_{i=1}^{SN} f_i t_i} \tag{9.2}$$

where fit$_i$ is the value of fitness for the minimization solution problems.

9.5.3 Scout Bee Phase

Trials numbers are majorly related to every source of food which has not further been updated or improved. If the food source is not upgraded by predefined tries by employed and onlooker bees, then the food source is considered to be deserted and the employed bees associated with the food source are transformed into the scout bees and then the scout bees' phase is initialized. Then, the source of food is replaced and the scout bee finds a new source of food. An essential control parameter for the predefined cycle's number is called the limit for rejection at ABC (Banharnsakun et al. 2010). At this time, scout bees are inserted into a new source of food.

9.5.4 Termination

The extinction criterion of ABC is commonly founded on an advanced number of groups or maximum cycle number (MCN) (Rao et al. 2008). This cycle number is presented by the worker prior to the simulation of the ABC algorithm.

The foraging bee's behavior has mainly four characteristics that defines self-organization and rely upon the following expressions.

  i. Positive feedback: If the nectar increases, the amount of food source increases, and visiting onlookers' bees can increase proportionally.
 ii. Negative feedback: Poor food source process exploitation was stopped by the bees.

iii. Fluctuations: Scout bees can transport the search procedure randomly for identification of the new sources of food.

iv. Several interactions: Food sources information is transferred by employed bees with nestmates (onlookers) which are waiting in the area of dance. The above re-examination is explained by the bees' foraging behavior, which fully satisfied the principle defined by Millonas (1993).

9.6 ABC in Process Optimization Methodology

The methodology for ABC optimization is commonly dependent on the concept of simulating the foraging behaviors of bees in the natural environment (El-Abd 2011; Hakli and Kiran 2020). Bees are naturally around their hives, exchanging and communicating information with each other on sources of food to maintain the populated colony. The two main steps of bees categorized above performed different tasks to discover the best source of food to sustain their colony (Figure 9.3). Onlooker and employed bees developed the source of food in their local neighborhood. Then, bees searched for solutions for the deterministic selection of employed bees and probabilistic selection of onlooker bees. Scout bees randomly attempt for a new source of food search for exploring the new regions for search space. These simulated bees then fly

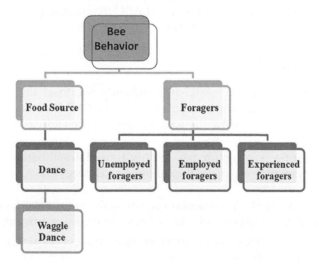

FIGURE 9.3
Steps involved in artificial bee colony methodology.

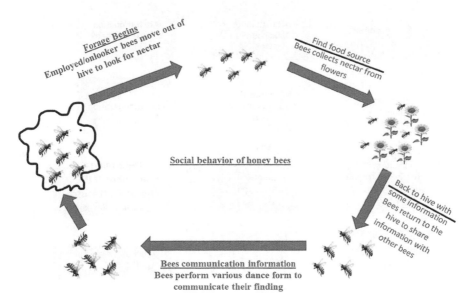

Social behavior of honey bees

FIGURE 9.4
Social behavior of honey bees.

around for the highly dimensional space search. In the ABC method, these bees combined the local search methodology for searching neighborhood sources of food; this process is known as the exploitation process, performed by the onlooker and employed bees, with the new region search through global food sources; this process is called exploration process (Babaeizadeh and Ahmad 2014), which is performed and managed by scout bees in an attempt to search for the best source of food for the colony.

During the food search, the larger number of honey bees are considered to be onlooker bees. To socially communicate with each other, waggle dance movement was performed (Figure 9.4). With this movement, scout bees update the employed honey bees about the superiority of the food source (Wang et al. 2010).

9.7 Novel Modified ABC (MABC)

A new modified artificial bee colony (MABC) is generally created using adaptive steps with exponential functions. This MABC is mainly used as an opposition-based theory for learning and an *S*-type improved method was used for grouping, and the MABC initial population is given for the original

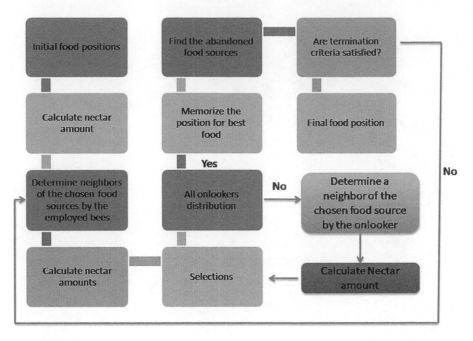

FIGURE 9.5
Different steps followed by a honey bee in ABC.

way of the roulette wheel sensitivity-pheromone selection (Figure 9.5 and Table 9.1). Particularly, an exponential adaptive step will design the functions for the replacement of the original step randomly, and the number of bees (onlooker and employed) are equal (Mohapatra et al. 2017; Kashan et al. 2012). This follows the many advantages of the ABC methodology such as simplicity, implementation ease, and performance that are also outstanding. There are, however, many flaws in the ABC algorithm: (i) The initial population generates the leading solution for a randomly dispersed population in the space. Therefore, the ability of searching is directly affected by it. (ii) All the bees begin to direct search in whole space solution, which may lead to a reduction of searching efficiency. (iii) Step length is randomly used for population regeneration. The search range should be usually restricted when the ABC performs in the neighborhood. Therefore, the optimization precision and the rate of convergence in the algorithm will be influenced (Kashan et al. 2012). Therefore, to overcome the problems with existing deficient algorithms and related ABCs, the MABC's purpose is to improve the

TABLE 9.1

Summary of ABC Algorithms' Recent Modifications

Algorithm Name	Description of Modifications	Problem	References
MABC	Incorporated Deb's rule in the selection of food sources employed and onlooker bees	Constrained Optimization	Karaboga et al. (2014)
ABC-AP	Adaptive penalty constraint and Deb's rule were used in the modification	Weight of truss structures	Sonmez (2011)
MABC	Control parameter introductions	Real-parameter optimization	Akay and Karaboga (2012)
IABC	Onlooker bee's movement was changed using universal gravitation force	Numerical optimization	Tsai et al. (2009)
ABCgBest and ABCgBestDist	Incorporated guest and distance-based reference selection movement of the onlooker	benchmark functions	Akdağli et al. (2011)
PABC-RC	Information about the integration exchange of ripple communication strategy	Benchmark functions	Liu et al. (2018)
ABC	For path search, Deb's rules were directly incorporated	Data clustering	Zhang et al. (2010)
ABC-PTS	Addition of new variables to the neighborhood search equation	Peak-to-average power ratio	Wang et al. (2020)
ABC	Addition of GRAH and two neighborhood structures for employed and onlooker bees	Generalized assignment	Baykaso et al. (2007)
ABC-PTS	Solution initialization transformation of ABC from continuous space vector into a discrete vector	Peak-to-average power ratio	Zhu and Kwong (2010)
DisABC	Use a differential expression to measure dissimilarity between binary vectors	Binary optimization	Karaboga and Basturk (2007)
ABC	Decision variables uses in indexes	Discrete optimum design of truss structures	Sonmez (2011)
OABC	Opposition concept number used	Black-Box Optimization Benchmarking (BBOB)	El-Abd (2011)

(Continued)

TABLE 9.1 (CONTINUED)

Algorithm Name	Description of Modifications	Problem	References
ES-ABC, MSABC, MH-ABC, and HH-ABC	ES-ABC local search, a master-slave technique in MS-ABC, multi-hive approach in MH-ABC, and combined MS-ABC and MH-ABC in HH-ABC were used	Numerical benchmark data	Gao et al. (2014)
binABC, normABC AMABC	Modified concept based on bin PSO and normalized DE, and angle-modulated PSO and DE	Binary optimization	Kashan et al. (2012)
DABC	Use of NM to start the population and local integrate search to ABC	Flow shop benchmark	Wang et al. (2020)

initial structure of the population, the consortium of the population, and the renewal of the ABC population. The MABC follows the following steps:

i. Initialization phase

CS/2 = Number for food, this is equivalent to half the colony size in the number of food sources for population initialization based on opposition learning

Fitness values calculations

Grouping of S-type sub-population

Zero = trials (Food Number 1). Trial should be zero

Initialized elitists of food sources

For memorization of the finest source of food, Iter = 0; while by sensitivity-pheromone Iter ≤ max iteration selects better elitist

Food number for employed bees' i = 1

Generate a new solution and update each food source step length

New solution evaluation

Iter = Iter + 1

The selection of sensitivity-pheromone was applied in between the present solution i and its other solution. Other solutions were applied if it is better than the present solution, i.e., the probability conditions.

ii. Onlooker bees

Food source selection by sensitivity-pheromone

For every onlooker bee, each source of food generates a new solution and updates the length of steps

Iter = Iter + 1

Selection applied to the sensitivity-pheromone to current solutions. But the existing resolution is better than the others. Otherwise, increasing its experimental counter-optimal solution should be memorized.

iii. Scout bee's phase

The food source determination of the trial counter should exceed the "limit" value, or in each interaction, only one scout bee was acceptable.

The scout bee used for random Food Number search ≥ max iteration returns to the optimal solution at the end.

9.8 Effect of Different Parameters in ABC of Enzyme Optimization

ABC algorithm should be more productive in terms of the scout bee's food source number (N), food source amount (M), elected food source number, dispatched bee's number to the elected food source (Nre), the bee's dispatched number to another source of food (Nsp), search area radius (Ngh), and iteration number (I_{max}) (Zhang et al. 2010). Initial food source locations with the conditions are represented within the well-defined problem for this algorithm.

$$X_{ij} = X_j^{min} + ran(0,1)\left(X_j^{min} - X_j^{min}\right) \tag{9.3}$$

where $i = 1, \ldots, N$ and $j = 1, \ldots, j$ are ranges in the equation. N is the number of parameters. In every area, subsequently, the new solutions created during ABC algorithm V_{jk} within X_k are demonstrated in Equation (9.4):

$$V_{jk}(t+1) = X_{jk}(t) + \varphi_{jk}(t)\left(X_{jk}(t) - X_{wk}(t)\right) \tag{9.4}$$

$$K = int(rand \times N) + 1$$

where $\varnothing_{jk}$ and X_{jk} represent the random numbers uniform distribution and the jth solution from among the solution set of the kth parameter. Though, the $\varnothing_{jk}$ area and parameter k are selected casually from the domains [1 and −1] and [1 and N], respectively. Beyond these situations, the explanation for each problem in a suitable way is replaced by the earlier one. The earlier solution is replaced if the new solution is more adaptable. After that, solutions for each bee are selected by the scout bees (Liu et al. 2018).

9.9 Control Parameters in ABC for Optimization

ABC optimization/adjustment of parameters are designed, after which the analysis was performed using the most appropriate design for the optimization of parameters. ABC algorithm can deliver the best parameters after the optimization of the process. The earlier ABC version was most efficient for basic functions. Although the performance of the ABC algorithm convergence was not that much effective when working with composite and non-separable constrained related problems and functions (Karaboga et al. 2014; Sonmez 2011). The convergence rate and improvement (Karaboga and Basturk 2007) were analyzed by the effect rate of perturbation which controls frequency change parameters; this method is commonly known as the scaling factor which determines the magnitude of parameters by changing neighboring solutions and the parameter limit by ABC performance; the modified version is proposed for solving efficient optimized problems for real parameter. Common modifications in the Artificial Bee Colony Algorithm (ABCA) are made in the perturbation process for controlling the frequency of perturbation. Frequency is fixed in the basic version of ABC while producing the new solution, with slow rate of convergence in only one parameter; this is the parent solution for the expected outcomes. But in the proposed ABCA (Mohapatra et al. 2017), a new control parameter – the modification rate (MR) – was introduced.

9.10 ABC Algorithm Modifications

Presents changes in the ABC subsection may cause the variations which are observed by different researchers in the relationships with modified tuning in the parameter, for enhancement or performance improvements. For tackling the constrained problems related to optimization, in the ABC selection procedure, the worker bees generally used Deb's rules to regulate the strategy in the procedure, instead of the greedy procedure for selection. The ABC's new variant performance was linked with two methods, i.e., PSO and DE, wherein the results are comparable and the ABC performance was clearly observed. For large-scale problems' solutions, the ABC is applied (Karaboga and Basturk 2007). The limited handling techniques were applied to the selection phase of ABC so that the feasible area of the whole search space could be reached. To choose the onlooker and employed bees, Deb's rules were incorporated. Three strategies were carried out for the selection and modification of the ABC. The modification in the three strategies selection in ABC was reported by Zhu and Kwong (2010) for numerical benchmark optimization. Modification in food source selection was carried out by onlooker bees for the prevention of premature convergence and population diversity increases. These strategies include selection based on rank selection-based selection (RABC), tournament selection (TABC), and disruptive selection of (DABC). Modified ABC performance is mainly associated with the results of basic and modified ABC as demonstrated by three selection strategies, and are used for population diversity improvement and convergence prevention.

9.11 Summary of ABC

1. Foraging behavior generally inspired by honey bees in ABC
2. For optimization purposes, ABC is used globally
3. Numerical optimization initially is proposed (Karaboga and Basturk 2007)
4. Combinatorial problems are analyzed for the optimization process (Pan et al. 2011)
5. It is also used for unconstrained and constrained problems related to optimization (Djaballah and Nouibat 2022; Karaboga et al. 2014; Kasihmuddin et al. 2021)
6. The user predetermines the employed control parameters (i.e., population size, number of maximum cycles, and limit)
7. It is comparatively simple, flexible, and robust (Rao et al. 2008; Singh 2009; Lin et al. 2021)

9.12 Application of ABC

The ABC has become the most common because of its ease of application and robustness. Researchers successfully applied the solution to complications in many different areas. First, the ABC applied to the mathematical problems by Karaboga et al. (2014) was extended to constrained optimization problems, and were then applied to neural networks of the brain (Karaboga and Basturk 2007), classification to the medical patterns, and clustering problems (Table 9.2) (Seyman and Taşpınar, 2013; Ilie and Bădică 2013).

TABLE 9.2

Applications of ABC in Different Fields

S. No.	Application of ABC	Reference
1	ABC is used for training the multilayer perceptron neural network for classifying the acoustic emission signal toward their particular source	Omkar and Senthilnath (2009)
2	Investigation of the comparison of the RBF neural network training algorithms in inertial sensor-based terrain classification	Kurban and Beşdok (2009)
3	In underbalanced drilling, they used ABC for the training of neural networks for bottom hole prediction of pressure	Irani and Nasimi (2011)
4	For modeling daily reference evapotranspiration (ET), they applied ABC in neural network training	Ozkan et al. (2011)
5	Used ABC to design the cloning template of goal-oriented C for cellular networks of neural architecture	Parmaksızoğlu Selami and Alçı Mustafa (2011)
6	Proposed an ABC-based methodology for the synthesis of neural networks	Garro et al. (2011)
7	They introduced the integrated system in which the wavelet transforms and neural network recruiting is majorly based on ABC stock forecasting	Hsieh and Yeh (2011)
8	They described a methodology based on ABC for maximizing the accuracy and minimizing the connection numbers for artificial neural networks through synaptic weights involvement	Kurban and Beşdok (2009)
9	ABC was used for MLP training and presented that the performance of MLP-ABC is improved than MLP-BP used for data of time series	Shah et al. (2012)
10	S-system neural networks were described as the basics of ABC neural network models for biochemical networks	Yeh and Hsieh (2012)

9.13 Conclusions and Future Prospects

The real-world complexity of the problems related to optimization increases the attractiveness of robust, fast, and accurate optimizers among scientists from different areas. Over the continuous and discrete spaces, ABC algorithm optimization is a newer and simpler population-based approach to optimization. It is a flexible and straightforward ABC algorithm that requires fewer parameters than the other algorithms. The original ABC algorithms and their modifications and hybrid algorithms are used for solving the optimization problems that are continuous, constrained, combinatorial, multiobjective, binary, chaotic, etc. Solving the various problems related to performed experiments from the literature proves that the ABC's efficiency, accuracy, and effectiveness are applied to all problems related to ABC. Reportedly, it has outperformed some EAs and other heuristic search tests for the benchmark problem of the real world. However, few control parameters are used and efficiently used to solve the problems related to multimodal and multidimensional optimization (Zhang et al. 2021). For the initial parameter's values, the ABC is not sensitive and is affected by increasing problems related to dimension, which is not similar to the other probabilistic algorithms related to optimization. ABC also has the inherent drawback of premature convergence and stagnation which adds to the capability loss of ABC in exploration and exploitation (Sharma et al. 2021). Although in the last two years, there have been many publications, there still exist many open complications and new areas of application where it should be useful, and also exist various dimensions in which this algorithm should be improved. Approximately, forthcoming critical research directions in the area of ABC are as follows. Complete search space investigation and management of region for optimal solution should be balanced by maintaining the diversity in the initial and future iterations for any arbitrary search algorithm for numbers. The updated equation for employed and onlooker bees phase in ABC should be as follows:

$$v_{ij} = A \times x_{ij} + B \times \left(x_{ij} - x_{kj} \right) \tag{9.5}$$

That is, modified position v_{ij} is the weighted sum of the food source position x_{ij} and the difference $(x_{ij} - x_{kj})$ of two positions of the source of food, where A is the mass of the target source of food and B is the mass difference of casual source of food. In a basic ABC, A is usually set to 1, while B is distributed uniformly in real numbers randomly (ϕ_{ij}) in the range [−1, 1]. Numerous studies were carried out on variations in ϕ_{ij} for a better mechanism of investigation (Karaboga et al. 2014). Moreover, for better outputs, the value and the range should also be fine-tuned. For the improvement in performance, effective modifications and implementation in the ABC should be done. It can be observed that many parts of the ABCA can be run in parallel. To estimate the performance and fitness, there is only one way, that is, parallel implementation should be there for every solution. Otherwise, the bees are distributed in the various processors which does not allows them improvement in the independent solution. Though this approach should be affected by the bee's dependencies in between, implementation parallel to the ABC should be considered for the shared memory of architecture, which overcomes these dependencies. The basic phase of onlooker bees in the ABC (Zhu and Kwong 2010) uses a roulette wheel selection scheme in which the fitness value of each slice is directly proportional to the size in selecting a suitable source of food. ABC is the multiobjective optimization for the solutions of the problems in which we have to optimize simultaneously two or more objectives (Banharnsakun et al. 2010). As the objective numbers increase, the Pareto-based conventional methods such as MOEAs may perform poorly. To solve this problem, the ABC multiobjective variants should be extended to research in features (Gao et al. 2014).

References

Agrawal Nikhil, Anil Kumar, and Varun Bajaj. 2016. "Optimized Design of Digital IIR Filter Using Artificial Bee Colony Algorithm." *Proceedings of 2015 International Conference on Signal Processing, Computing and Control, ISPCC 2015*, no. 1: 316–21. doi:10.1109/ISPCC.2015.7375048.

Akay, Bahriye, and Dervis Karaboga. 2012. "A Modified Artificial Bee Colony Algorithm for Real-Parameter Optimization." *Information Sciences* 192. Elsevier Inc.: 120–42. doi:10.1016/j.ins.2010.07.015.

Akdağli, Ali, Mustafa Berkan Biçer, and Seda Ermiş. 2011. "A Novel Expression for Resonant Length Obtained by Using Artificial Bee Colony Algorithm in Calculating Resonant Frequency of C-Shaped Compact Microstrip Antennas." *Turkish Journal of Electrical Engineering and Computer Sciences* 19 (4): 597–606. doi:10.3906/elk-1006-466.

Babaeizadeh, Soudeh, and Rohanin Ahmad. 2014. "An Efficient Artificial Bee Colony Algorithm for Constrained Optimization Problems." *Journal of Engineering and Applied Sciences* 9 (10): 405–13. doi:10.3923/jeasci.2014.405.413.

Babaoglu, Ismail. 2015. "Artificial Bee Colony Algorithm with Distribution-Based Update Rule." *Applied Soft Computing Journal* 34. Elsevier B.V.: 851–61. doi:10.1016/j.asoc.2015.05.041.

Banharnsakun, Anan, Tiranee Achalakul, and Booncharoen Sirinaovakul. 2010. "Artificial Bee Colony Algorithm on Distributed Environments." *Computer Engineering*, 13–18.

Bansal, Jagdish Chand, Harish Sharma, and K. v. Arya. 2011. "Model Order Reduction of Single Input Single Output Systems Using Artificial Bee Colony Optimization Algorithm." *Studies in Computational Intelligence* 387: 85–100. doi:10.1007/978-3-642-24094-2_6.

Baykaso, Adil, Lale Özbakır, and Pınar Tapkan. 2007. *Artificial Bee Colony Algorithm and Its Application to Generalized Assignment Problem. Swarm Intelligence: Focus on Ant and Particle Swarm Optimization.*

Beg, Mirza Samiulla, Akhilesh A. Waoo, Ritu Gautam, Prableen Kaur, Manik Sharma, Yan Song, Lidong Huang, et al. 2012. "Improved Binary PSO for Feature Selection Using Gene Expression Data." *Computational Biology and Chemistry* 32 (1). Elsevier B.V.: 29–38. doi:10.1016/j.compbiolchem.2007.09.005.

Brezočnik, Lucija, Iztok Fister, and Vili Podgorelec. 2018. "Swarm Intelligence Algorithms for Feature Selection: A Review." *Applied Sciences (Switzerland)* 8 (9). doi:10.3390/app8091521.

Chang, Po Chun, and Xiangjian He. 2014. "Macroscopic Indeterminacy Swarm Optimization (MISO) Algorithm for Real-Parameter Search." *Proceedings of the*

2014 IEEE Congress on Evolutionary Computation, CEC 2014, 1571–78. doi:10.1109/ CEC.2014.6900281.

Chuang, Li Yeh, Hsueh Wei Chang, Chung Jui Tu, and Cheng Hong Yang. 2008. "Improved Binary PSO for Feature Selection Using Gene Expression Data." *Computational Biology and Chemistry* 32 (1): 29–38. doi:10.1016/j. compbiolchem.2007.09.005.

Djaballah, Chouaib Ben, and Wahid Nouibat. 2022. "A New Multi-Population Artificial Bee Algorithm Based on Global and Local Optima for Numerical Optimization." *Cluster Computing*. doi:10.1007/s10586-021-03507-w.

El-Abd, Mohammed. 2011. "Opposition-Based Artificial Bee Colony Algorithm." *Genetic and Evolutionary Computation Conference, GECCO'11*, no. 1: 109–15. doi:10.1145/2001576.2001592.

Gao, Wei Feng, San Yang Liu, and Ling Ling Huang. 2014. "Enhancing Artificial Bee Colony Algorithm Using More Information-Based Search Equations." *Information Sciences* 270. Elsevier Inc.: 112–33. doi:10.1016/j.ins.2014.02.104.

Garlapati, Vijay Kumar, and Rintu Banerjee. 2010. "Evolutionary and Swarm Intelligence-Based Approaches for Optimization of Lipase Extraction from Fermented Broth." *Engineering in Life Sciences* 10 (3): 265–73. doi:10.1002/ elsc.200900086.

Garoudja, Elyes, Kamel Kara, Aissa Chouder, and Santiago Silvestre. 2015. "Parameters Extraction of Photovoltaic Module for Long-Term Prediction Using Artifical Bee Colony Optimization." *3rd International Conference on Control, Engineering and Information Technology, CEIT 2015*. doi:10.1109/CEIT.2015.7232993.

Garro, Beatriz A., Humberto Sossa, and Roberto A. Vazquez. 2011. "Artificial Neural Network Synthesis by Means of Artificial Bee Colony (ABC) Algorithm." *2011 IEEE Congress of Evolutionary Computation, CEC 2011*, May 2014: 331–38. doi:10.1109/CEC.2011.5949637.

Gautam, Ritu, Prableen Kaur, and Manik Sharma. 2019. "A Comprehensive Review on Nature Inspired Computing Algorithms for the Diagnosis of Chronic Disorders in Human Beings." *Progress in Artificial Intelligence* 8 (4). Springer Berlin Heidelberg: 401–24. doi:10.1007/s13748-019-00191-1.

Hakli, Huseyin, and Mustafa Servet Kiran. 2020. "An Improved Artificial Bee Colony Algorithm for Balancing Local and Global Search Behaviors in Continuous Optimization." *International Journal of Machine Learning and Cybernetics* 11 (9). Springer Berlin Heidelberg: 2051–76. doi:10.1007/s13042-020-01094-7.

Harrison, Kyle Robert, Andries P. Engelbrecht, and Beatrice M. Ombuki-Berman. 2018. *Self-Adaptive Particle Swarm Optimization: A Review and Analysis of Convergence. Swarm Intelligence.* Vol. 12. Springer US. doi:10.1007/s11721-017-0150-9.

Hsieh, Tsung Jung, and Wei Chang Yeh. 2011. "Knowledge Discovery Employing Grid Scheme Least Squares Support Vector Machines Based on Orthogonal Design Bee Colony Algorithm." *IEEE Transactions on Systems, Man, and Cybernetics, Part B: Cybernetics* 41 (5): 1198–1212. doi:10.1109/TSMCB.2011.2116007.

Hu, Yu, Zhensheng Sun, Lijia Cao, Yin Zhang, and Pengfei Pan. 2021. "Optimization Configuration of Gas Path Sensors Using a Hybrid Method Based on Tabu Search Artificial Bee Colony and Improved Genetic Algorithm in Turbofan Engine." *Aerospace Science and Technology* 112. doi:10.1016/j.ast.2021.106642.

Ilie, Sorin, and Costin Bădică. 2013. "Multi-Agent Distributed Framework for Swarm Intelligence." *Procedia Computer Science* 18: 611–20. doi:10.1016/j. procs.2013.05.225.

Irani, Rasoul, and Reza Nasimi. 2011. "Application of Artificial Bee Colony-Based Neural Network in Bottom Hole Pressure Prediction in Underbalanced Drilling." *Journal of Petroleum Science and Engineering* 78 (1). Elsevier B.V.: 6–12. doi:10.1016/j.petrol.2011.05.006.

Karaboga, Dervis, and Bahriye Basturk. 2007. "A Powerful and Efficient Algorithm for Numerical Function Optimization: Artificial Bee Colony (ABC) Algorithm." *Journal of Global Optimization* 39 (3): 459–71. doi:10.1007/s10898-007-9149-x.

Karaboga, Dervis, Beyza Gorkemli, Celal Ozturk, and Nurhan Karaboga. 2014. "A Comprehensive Survey: Artificial Bee Colony (ABC) Algorithm and Applications." *Artificial Intelligence Review* 42 (1): 21–57. doi:10.1007/s10462-012-9328-0.

Kashan, Mina Husseinzadeh, Nasim Nahavandi, and Ali Husseinzadeh Kashan. 2012. "DisABC: A New Artificial Bee Colony Algorithm for Binary Optimization." *Applied Soft Computing Journal* 12 (1). Elsevier B.V.: 342–52. doi:10.1016/j.asoc.2011.08.038.

Kasihmuddin, Mohd Shareduwan Mohd, Mohd Asyraf Mansor, Shehab Abdulhabib Alzaeemi, and Saratha Sathasivam. 2021. "Satisfiability Logic Analysis via Radial Basis Function Neural Network with Artificial Bee Colony Algorithm." *International Journal of Interactive Multimedia and Artificial Intelligence* 6 (6): 164–73. doi:10.9781/ijimai.2020.06.002.

Kurban, Tuba, and Erkan Beşdok. 2009. "A Comparison of RBF Neural Network Training Algorithms for Inertial Sensor Based Terrain Classification." *Sensors* 9 (8): 6312–29. doi:10.3390/s90806312.

Leu, Min Shyang, and Ming Feng Yeh. 2012. "Grey Particle Swarm Optimization." *Applied Soft Computing Journal* 12 (9). Elsevier B.V.: 2985–96. doi:10.1016/j asoc.2012.04.030.

Li, Bai, Mu Lin, Qiao Liu, Ya Li, and Changjun Zhou. 2015. "Protein Folding Optimization Based on 3D Off-Lattice Model via an Improved Artificial Bee Colony Algorithm." *Journal of Molecular Modeling* 21 (10). doi:10.1007/s00894-015-2806-y.

Li, Nailu, Hua Yang, and Anle Mu. 2019. "Improved Grey Particle Swarm Optimization and New Luus-Jaakola Hybrid Algorithm Optimized IMC-PID Controller for Diverse Wing Vibration Systems." *Complexity* 2019. doi:10.1155/2019/8283178.

Lin, Yanhong, Jing Wang, Xiaolin Li, Yuanzi Zhang, and Shiguo Huang. 2021. "An Improved Artificial Bee Colony for Feature Selection in QSAR." *Algorithms* 14 (4). doi:10.3390/a14040120.

Liu, Jiangshan, Yumei Zhang, and Shulin Bai. 2018. "An Adaptive Artifical Bee Colony Algorithms Based on Global Best Guide." *Proceedings – 13th International Conference on Computational Intelligence and Security, CIS 2017* 2018-Janua: 211–15. doi:10.1109/CIS.2017.00053.

Liu, Yi, and Shengqing Tang. 2018. "An Application of Artificial Bee Colony Optimization to Image Edge Detection." *ICNC-FSKD 2017 – 13th International Conference on Natural Computation, Fuzzy Systems and Knowledge Discovery*. IEEE, 923–29. doi:10.1109/FSKD.2017.8393400.

Mahapatra, Paramita, Annapurna Kumari, Vijay Kumar Garlapati, Rintu Banerjee, and Ahindra Nag. 2009. "Enzymatic Synthesis of Fruit Flavor Esters by Immobilized Lipase from Rhizopus Oligosporus Optimized with Response

Surface Methodology." *Journal of Molecular Catalysis B: Enzymatic* 60 (1–2): 57–63. doi:10.1016/j.molcatb.2009.03.010.

Millonas, Mark M. 1993. "'Swarms, Phase Transitions, and Collective Intelligence.'" *ArXiv:* Adaptation and Self-Organizing Systems.

Mohapatra, Prabhujit, Kedar Nath Das, and Santanu Roy. 2017. "A Modified Competitive Swarm Optimizer for Large Scale Optimization Problems." *Applied Soft Computing Journal* 59. Elsevier B.V.: 340–62. doi:10.1016/j.asoc.2017.05.060.

Omkar, S.N., and J. Senthilnath. 2009. "Artificial Bee Colony for Classification of Acoustic Emission Signal Source." *International Journal of Aerospace Innovations* 1 (3): 129–43. doi:10.1260/175722509789685865.

Ozkan, Coskun, Ozgur Kisi, and Bahriye Akay. 2011. "Neural Networks with Artificial Bee Colony Algorithm for Modeling Daily Reference Evapotranspiration." *Irrigation Science* 29 (6): 431–41. doi:10.1007/s00271-010-0254-0.

Pan, Quan Ke, M. Fatih Tasgetiren, P. N. Suganthan, and T. J. Chua. 2011. "A Discrete Artificial Bee Colony Algorithm for the Lot-Streaming Flow Shop Scheduling Problem." *Information Sciences* 181 (12). Elsevier Inc.: 2455–68. doi:10.1016/j.ins.2009.12.025.

Rao, R. Srinivasa, S. V. L. Narasimham, and M. Ramalingaraju. 2008. "Optimization of Distribution Network Configuration for Loss Reduction Using Artificial Bee Colony Algorithm." *International Journal of Electrical and Computer Engineering* 2 (9): 1964–70. doi:10.5281/zenodo.1057591.

Sathesh Kumar, K., and M. Hemalatha. 2014. "An Innovative Potential on Rule Optimization Using Fuzzy Artificial Bee Colony." *Research Journal of Applied Sciences, Engineering and Technology* 7 (13): 2627–33. doi:10.19026/rjaset.7.578.

Selami, Parmaksızoğlu and Alçı Mustafa. 2011. "A Novel Cloning Template Designing Method by Using an Artificial Bee Colony Algorithm for Edge Detection of CNN Based Imaging Sensors." *Sensors* 11(5): 5337–5359. doi:10.3390/s110505337

Seyman, M. N., and N. Taşpınar. 2013. "Pilot Tones Optimization Using Artificial Bee Colony Algorithm for MIMO-OFDM Systems." *Wireless Personal Communications* 71(1): 151–163. https://doi.org/10.1007/s11277-012-0807-z

Shah, Habib, Rozaida Ghazali, and Nazri Mohd Nawi. 2012. Hybrid Ant Bee Colony Algorithm for Volcano Temperature Prediction. In: Chowdhry, B.S., Shaikh, F.K., Hussain, D.M.A., Uqaili, M.A. (eds) *Emerging Trends and Applications in Information Communication Technologies.* IMTIC 2012. Communications in Computer and Information Science, vol 281. Springer, Berlin, Heidelberg. https://doi.org/10.1007/978-3-642-28962-0_43

Sharma, Kavita, P. C. Gupta, and Nirmala Sharma. 2021. "Limaçon Inspired Artificial Bee Colony Algorithm for Numerical Optimization." *Evolutionary Intelligence* 14 (3). Springer Berlin Heidelberg: 1345–53. doi:10.1007/s12065-020-00430-8.

Singh, Alok. 2009. "An Artificial Bee Colony Algorithm for the Leaf-Constrained Minimum Spanning Tree Problem." *Applied Soft Computing Journal* 9 (2): 625–31. doi:10.1016/j.asoc.2008.09.001.

Sonmez, Mustafa. 2011. "Artificial Bee Colony Algorithm for Optimization of Truss Structures." *Applied Soft Computing Journal* 11 (2). Elsevier B.V.: 2406–18. doi:10.1016/j.asoc.2010.09.003.

Tsai, Pei Wei, Jeng Shyang Pan, Bin Yih Liao, and Shu Chuan Chu. 2009. "Enhanced Artificial Bee Colony Optimization." *International Journal of Innovative Computing, Information and Control* 5 (12): 5081–92. doi:10.1002/elps.200900194.

Wang, Juan, Xu Honglei, Kok Lay Teo, Jie Sun, and Jianxiong Ye. 2020. "Mixed-Integer Minimax Dynamic Optimization for Structure Identification of Glycerol Metabolic Network." *Applied Mathematical Modelling* 82. Elsevier Inc.: 503–20. doi:10.1016/j.apm.2020.01.042.

Wang, Yajun, Wen Chen, and Chintha Tellambura. 2010. "A PAPR Reduction Method Based on Artificial Bee Colony Algorithm for OFDM Signals." *IEEE Transactions on Wireless Communications* 9 (10): 2994–99. doi:10.1109/TWC.2010.081610.100047.

Xiang, Wan Li, and Mei Qing An. 2013. "An Efficient and Robust Artificial Bee Colony Algorithm for Numerical Optimization." *Computers and Operations Research* 40 (5). Elsevier: 1256–65. doi:10.1016/j.cor.2012.12.006.

Yeh, Wei Chang, and Tsung Jung Hsieh. 2012. "Artificial Bee Colony Algorithm-Neural Networks for S-System Models of Biochemical Networks Approximation." *Neural Computing and Applications* 21 (2): 365–75. doi:10.1007/s00521-010-0435-z.

Zhang, Changsheng, Dantong Ouyang, and Jiaxu Ning. 2010. "An Artificial Bee Colony Approach for Clustering." *Expert Systems with Applications* 37 (7). Elsevier Ltd: 4761–67. doi:10.1016/j.eswa.2009.11.003.

Zhang, Jianchun, Lei Li, and Zhiwei Chen. 2021. "Strength–Redundancy Allocation Problem Using Artificial Bee Colony Algorithm for Multi-State Systems." *Reliability Engineering and System Safety* 209 (September 2020). Elsevier Ltd: 107494. doi:10.1016/j.ress.2021.107494.

Zhang, Xiangrong, Wenna Wang, Yangyang Li, and L. C. Jiao. 2012. "PSO-Based Automatic Relevance Determination and Feature Selection System for Hyperspectral Image Classification." *Electronics Letters* 48 (20): 1263–65. doi:10.1049/el.2012.0539.

Zhu, Guopu, and Sam Kwong. 2010. "Gbest-Guided Artificial Bee Colony Algorithm for Numerical Function Optimization." *Applied Mathematics and Computation* 217 (7). Elsevier Inc.: 3166–73. doi:10.1016/j.amc.2010.08.049.

Index